BIRMINGHAM CITY
UNIVERSITY
DISCARDED

Contaminated Land

To those still carrying on...

Aaron, Aidan, Aviva, Bernadette, Bertha, Derek, Fred, Hannah, Helen, Justin, Lane, Laura, Lewis, Marcy, Nina

...and those yet to come that have to live with it all.

Contaminated Land

Reclamation, Redevelopment and Reuse in the United States and the European Union

Peter B. Meyer
Professor of Urban Policy and Economics and Director, Center for Environmental Management
University of Louisville, US

Richard H. Williams
Associate Centre for Research on European Urban Environments and Lecturer in Town and Country Planning
University of Newcastle, UK

Kristen R. Yount
Assistant Professor of Sociology
Northern Kentucky University, US

Edward Elgar
Aldershot, UK • Brookfield, US

© Peter B. Meyer, Richard H. Williams and Kristen R. Yount 1995

All rights reserved. No part of this publication may be reproduced, stored in a retrieval system or transmitted in any form or by any means, electronic, mechanical, photocopying, recording, or otherwise without the prior permission of the publisher.

Published by
Edward Elgar Publishing Limited
Gower House
Croft Road
Aldershot
Hants GU11 3HR
UK

Edward Elgar Publishing Company
Old Post Road
Brookfield
Vermont 05036
US

British Library Cataloguing in Publication Data

Meyer, Peter B.
    Contaminated Land: Reclamation,
    Redevelopment and Re-use in the United
    States and the European Union
    I. Title
    333.73153

Library of Congress Cataloguing in Publication Data

Meyer, Peter B., 1943–
    Contaminated land: reclamation, redevelopment and re-use in the United States and the European Union / Peter B. Meyer, Richard H. Williams, Kristen R. Yount.
        1. Soil pollution—Government policy—United States. 2. United States. Comprehensive Environmental Response, Compensation, and Liability Act of 1980. 3. Soil remediation—United States. 4. Soil pollution—Government policy—European Union countries. 5. Soil remediation—European Union countries. I. Williams, Richard H., 1949– . II. Yount, Kristen R. III. Title
TD878.M48 1996
363.73'968'0973—dc20   95-19495
                                                                                         CIP

ISBN 1 85898 243 X

Printed in Great Britain at the University Press, Cambridge

Contents

Tables and Figures vii
Preface and Acknowledgements viii

PART I

1. Introduction: The Contaminated Land Problem 3
   What is the Problem? 5
   Contaminated Land Politics, Policy and Planning 6
   Comparative Analysis and our Logic of Argument 11
   The Extent of the Contaminated Land Problem 13
   Who is Hurt by Contamination and Cleanup Costs and
     Delays? 15
   Costs Associated with Contaminated Land Cleanup Efforts 19
2. Dimensions of National Contaminated Land Policy Contexts 24
   Past Environmental Experiences: Accidents, Exposures and
     'Disasters' 25
   The State Politico-Legal Context 27
   The Locus of Control over the Use of Land 32
   Approaches to Liability for Damages and Cleanups 36
   Access to Capital and to Liability Insurance 44
   Current Policy and Political Pressures 49

PART II

3. The Emergence of Environmental Concern and Toxics
     Policy in the United States 55
   The Emergence of US Environmentalism 55
   The Environmental Protection Agency 58
   The Regulation of Hazardous Substances and Wastes 60
4. The CERCLA Experience and Debates over Change 71
   National Priority List Cleanups 72
   Financing the Fund 76
   Liability Provisions under CERCLA: Equity Issues 77
   The Impact of CERCLA on Brownfield Redevelopment 84

State Legislation to 'Solve' the Redevelopment Problem	91
Recent CERCLA Reauthorization and Change Proposals	95
Examples of CERCLA Successes	98

PART III

5. The European Context and European Union Environmental
   - Policy — 103
   - Which 'Europe'? — 104
   - European Union Environmental Control Powers — 109
   - Implementation of the EU Environmental Programme — 115
   - The Rationale for an EU Contaminated Land Policy — 119
6. Contaminated Land Policy in the European Union and its
   - Member-States — 121
   - The Contamianted Land Problem — 122
   - Contaminated Land Policy Development in the EU — 127
   - Issues in the Ongoing EU Contaminated Land Policy Debate — 129
   - Diversity in Member-States' Conditions and Policies — 137

PART IV

7. Comparisons and Contrasts: Integrated Comparative Analysis — 147
   - Past Environmental Experiences: Accidents, Exposures and 'Disasters' — 147
   - The State Politico-Legal Context — 149
   - The Locus of Control over the Use of Land — 151
   - Approaches to Liability for Damages and Cleanup — 152
   - Access to Capital and to Liability Insurance — 159
   - Current Policy and Political Pressures — 165
   - The Competitive Position of Contaminated Lands — 169
8. Lessons for Future Contaminated Land Policy: Prospects and Pitfalls — 175
   - The Experience to Date: A Summary Assessment — 176
   - Principles for Managing the Urban Common Pool Resource — 178
   - Elements of an Effective Contaminated Land Policy — 181
   - Varieites of Risk Exposure — 187
   - The Pressures for Policy Changes — 194
   - Conclusions: What Works? — 199

References	203
Index	218

Tables and Figures

List of Tables

1.1	The extent of land contamination	15
5.1	Selected events in the evolution of EU environmental policy	105
7.1	The state politico-legal context	150
7.2	The locus of control over the use of land	153
7.3	Approaches to liability for damages and cleanup	154
7.4	Access to capital and to liability insurance	163
7.5	Investors' costs of contaminated land policies	170

List of Figures

2.1	Dimensions of the contaminated land policy context	26

Preface and Acknowledgements

This book has its origins in concerns over economic development and environmental justice. It was shaped by the processes of interdisciplinary discourse and refined by the complexities of cross-cultural and cross-national comparisons.

Two of us, Peter Meyer and Kristen Yount, an economist and sociologist, began work in 1992 on the problems posed for economic revitalization in US cities by the Comprehensive Environmental Liability, Compensation and Reclamation Act of 1970. While we discuss problems of language and shared understanding in the context of European Union environmental and planning policy in this book, we first had to overcome the problems of disparate meanings attached to common words in discourse between an economist and a sociologist. Simultaneously, we had to reach beyond our backgrounds in different subsets of our disciplines to encompass a vastly greater body of thought and literature, from real estate development and investment finance through the social psychology of risk and risk behaviours of organizations, and on to questions involving risk communication and the effects of uncertainty on risk perception and response.

While Kristen and Peter were observing the US scene, Dick Williams, the planner in our interdisciplinary triad and the European counterbalance to our two members from the United States, was engaged in a series of studies of the real estate markets in different member-states of the European Union contemporaneously with his work on policy formation and planning processes in the nations of the new Europe. He dealt more directly with the problem of language and communication, engaging in extensive work on planning and environmental policy in Germany, especially after reunification, when he encountered most directly the problems of extreme land contamination.

The trans-Atlantic bridge was anchored by Peter, who has been writing about Europe (primarily Britain) and the United States for the better part of a decade. He presented a problem for his co-authors, however, since he could not provide critical reading of their work, having no sense of what would or would not be comprehensible to readers from one side of the Atlantic reading about the other. Kristen and Dick, therefore, had to help each other communicate the nuances of political climate, social norms and legal systems, as well as overcome the problem George Bernard Shaw noted when he declared the British and the Americans to be 'two peoples separated by a common language'.

Our disparate experiences shaped this volume, as did the intellectual and political environment in which we were writing. The taxing authority for the key 'Superfund' under the US law expires at the end of 1995, and reauthorization legislation was needed for its continuation — but the anti-regulatory rhetoric and priorities of the new Republican Congress suggested it might be permitted to die out. The controversy produced myriad proposals for reform and revision of the law which certainly enriched our perspective and provided empirical findings on which we have drawn. The European context, while not convulsed by changes in party control of the legislative body which had not occurred in 30 years, was also fraught with turmoil: 1994 was a critical period for the new European Union, and the overdue European Environmental Agency was finally implemented, while individual member-states have struggled with increasingly complex and uncertain legislative approaches to land contamination within their borders.

Like any such policy-relevant book, our work has been shaped by its context. We contend that the principles derived, and the methods for comparison we illustrate, have a longer life and constitute a broader contribution. Our examples, and the policy choices we consider, reflect the current political debates, but the logic we employ, and the derivation of the social, political and legal contexts in which land use and related regulations need to be considered and compared, should, we hope, prove to be more lasting insights, applicable to the new and different land use and environmental policy decisions that will inevitably face the industrial democracies on both sides of the Atlantic.

We owe a number of debts of gratitude to an array of parties that contributed their insights and in some cases, their resources to our research. We can 'blame' Patsy Healey of the University of Newcastle for bringing us together. She helped link us up, but bears no responsibility for the result of that tie. Immediate intellectual debts are due to Simin Davoudi, Colin Wymer, Michaele Aden, Dick Bingham, Peter Noll, Charlie Bartsch, Hartmut Dieterich, Nancey Green Leigh, Martin Stein, Chris Walker and many others who share responsibility for shaping our thoughts on contaminated lands — but not for the conclusions we have offered on paper. We do not dare to trace all our intellectual debts, for the list of contributors would be too long.

The expensive processes of cross-Atlantic collaboration as well as a great deal of the work done in the United States was facilitated by support funds from the Institute for the Environment and Sustainable Development of the University of Louisville. Peter's sabbatical leave permitted him to pull the pieces of the book together in Fall 1994, so we owe an exceptional debt of gratitude to that institution's Doctoral Program in Urban and Public Affairs. The University of Newcastle Department of Town and Country Planning and its Centre for Research in European Urban Environments provided the facilities permitting Dick to make his contributions to our collaboration while the University of Northern Kentucky Department of Sociology, Anthropology and Philosophy did the same for Kristen, including providing her with a one-course teaching load reduction in Fall 1994.

We offer anonymous thanks to developers, bankers, legislators, engineers, regulators, surveyors and other professionals dealing with contaminated lands who helped us learn the details of their experiences and concerns in making decisions about actions to take on such properties. While we expect them to prefer their anonymity, we can thank in person Martin Kelly of Triumvirate Risk Management, Ltd., London, and Chris Walker of the Urban Institute and Charlie Bartsch of the Northeast-Midwest Institute, both in Washington, DC, for providing key access to individuals and documents that have made substantial contributions to this study. To credit Angela Keene for manuscript preparation is to understate her contribution. Skilled computer jockey, ex-typesetter, and superior word processor, she *literally* 'made this book' possible.

Many have contributed and many have guided, but our filters have produced the final picture. For that, we alone take responsibility.

PART I

1. Introduction: The Contaminated Land Problem

Over one hundred and fifty years of industrial development in the wealthiest nations in the world have left their mark. Beginning shortly after the completion of rebuilding efforts following World War II, attention to land reclamation and redevelopment shifted from an exclusive focus on bombed-out areas to the many abandoned factories, rail yards, shipyards and mines left behind by advancing economies and technologies — and to the residential areas near them that had housed the workers of the industrial revolution. At first, the concerns for these areas were land underutilization, regional economic decline and the need for new capital for revitalization and new job and income creation. Country after country turned to national efforts to modify tendencies towards the use and abandonment of land and communities by ever more mobile private sector capital.

The United States created a number of different programmes during the Depression in the 1930s to reclaim abandoned sites and put people to work, including branches of the Civilian Conservation Corps and of the Works Progress Administration. While federal programmes to promote new housing construction exploded in volume after World War II, it was not until 1965 that the Department of Housing and Urban Development was created to address the special problems of urban centres, especially those associated with capital flight. The US eventually promulgated its Economic Development Administration, an office in the far older Department of Commerce designed to assist localities abandoned by mobile capital, in 1965, despite a staunch refusal to adopt systematic compensatory regional policies as a matter of national policy.

Britain first launched a form of regional policy with its Special Areas Act of 1934, and has experimented with many different regional policy instruments since then. Linkage with the specific aim of reusing previously developed sites was not evident before the 1960s, becoming explicit in many urban regeneration initiatives in the following decade. Formal spatially-focused economic reconstruction and redevelopment programmes have, of course, characterized all the nations that suffered major impacts from World War II. Recycling of brownfield land was a major component of France's 1984 regional policy and growth pole strategy. Property redevelopment and avoidance of any underutilization or abandonment of land has been integral to land-use planning doctrine in The Netherlands since early in the inter-war period.

Thus, for over fifty years, some economic development officials and organizations concerned with attracting capital to particular regions recognized a need to distinguish between 'brownfield' and 'greenfield' sites, although the labels entered the language far later. 'Brownfields' were developed sites, with buildings and facilities from earlier industrial periods that often could not serve new types of businesses. Other things being equal, they were economically inferior to their greenfield competition as new production facilities. They required building removal and land clearing, possibly the acquisition of many smaller plots to form a single large site for modern single-storey production facilities, and otherwise engendered costs for *re*development not present in sites that had not previously been used for industrial or commercial purposes.

This relative disadvantage has been made more severe by the growth in environmental concerns in recent years. The recognized residue of prior economic activity broadened from merely buildings that needed to be cleared for new uses to include chemicals stored on the properties and spilled into the soil. Thus older industrial areas — often major portions of the land areas of historically significant urban centres that continued to house large proportions of national populations — faced growing problems in attracting the new investment and development capital essential to sustaining income and jobs in increasingly global economies. Whatever the biologically determinable ecological or human health impacts of the chemical leftovers from earlier productive activities, such 'contaminated lands' engendered growing economic, social, and, ultimately, political problems.

Derelict land and brownfields were of concern to European countries well before the US recognized the problem at the national level (apparently because of its ample land supply). The US led the way with legislation addressing the specific problem of *contaminated* brownfields in 1980, followed by gradual inclusion of more environmental regulations into Britain's Derelict Land Grant programme, which dated back to 1966. By contrast, the countries of continental Europe, while they have addressed individual cases of

Introduction: The Contaminated Land Problem

contamination, have not promulgated comprehensive national legislation addressing the problem comprehensively. No coherent and consistent approach has emerged — and, worse yet, no good data are available nor evaluations completed of the effects of any of the various laws and regulations on cleaning up past pollution and/or stimulating reuse of brownfield sites. We have contaminated lands — and policies towards them — but no evidence of solutions to 'the Contaminated Land Problem'.

WHAT IS THE PROBLEM?

Arguably, all residents of the nations that led the industrial revolution have benefited from the past environmental neglect that has contributed to their current high levels of per capita income. In effect, these economies borrowed from their futures by not cleaning up after themselves. The debts thus incurred were in many instances not voluntary or conscious: the dangers associated with indiscriminate disposal of chemicals used in production were not known or understood in most instances. With the growth of concern over those effects, however, the loans are now being called, and the debt must be repaid with interest. Repayment takes the form of cleanup and safe disposal of the chemical and other toxic residues of past production practices.

The problem has been stated cogently by the retired Director of the US Environmental Protection Agency's office of Toxic Substances and Environmental Monitoring Systems laboratory:

> Americans have no choice but to pay now or pay later for their long term survival. Actions or inactions during the 1990s will determine the costs during the next century. The monumental debt is accumulating, and the price tag for healing America's chemical wounds increases every year. (Schweitzer 1991:280)

The contaminated land issue may thus be bisected into: first, the magnitude of this accumulated debt, and second, the manner in which it will be paid, most specifically, what portions of the obligation to clean up contamination will be paid by which economic entities or participants in national economies. It is clear that, with respect to the first issue, the magnitude of the debt, biological and chemical scientific findings are central to determination of toxicity and threats to ecosystems and human health. However, whatever the scientific findings, the public policy problem, our primary concern here, lies in the politically-defined extent of cleanup required.

The financial costs of such cleanups are the accumulated obligation. These costs are affected by both the standards employed to define cleanliness and the technologies applied for mitigation of pollution and its damage. To the extent that discharge of this debt is a matter of public policy, the standards and acceptable technologies evolve from a political process. Scientific findings on the extent of contamination and the risks it poses are filtered through public debate. Standards can vary from one country to another, or change within a country as opinions and political imperative shift over time. The contaminated land problem is thus both a matter of how much needs to be paid as well as who should pay.

CONTAMINATED LAND POLITICS, POLICY AND PLANNING

We, the authors, are social scientists and planners; we do not presume to contribute to the scientific debate on the 'true' magnitude of the debt for past pollution. We note that 'truth' is not immutable, and scientific 'facts' regarding the extent of contamination, its severity, and the projected costs of cleanup continue to change over time. The changing policy concern over contaminated land reflects just such a moving scientific base: the extent of pollution and its effects continue to be better understood as techniques for measurement and comprehension of ecological processes develop over time. The very absence of rigid standards is, in fact, a major issue in the financial community's concern about unpredictable potential liabilities associated with land contamination and their effects on the profits from redevelopment investments.

Our objective here is to provide a framework for comparison and to determine the factors shaping the impact of contaminated land policies on brownfield redevelopment in different contexts. Policy 'success' may be measured by some combination of cleanup or reclamation rates, levels of redevelopment and/or rates of reuse of brownfield sites, where such data are available. To the extent that we can assess effects on levels of reclamation, redevelopment and reuse in different settings, we may be able to generalize the factors that contribute to the mitigation of land contamination. We thus examine the contexts in which the contaminated land problem is identified and 'resolved' in different

Introduction: The Contaminated Land Problem

countries. Such a comparative analysis of the political and cultural forces that shape contaminated land policy is essential to identifying the planning factors and outcomes that contribute to success.

To address the question of the severity of the contamination problem, we examine political processes and institutions, and cultural values and traditions, and how they shape the recognition and definition of what a society and nation-state sees as its *effective* debt. This obligation is the contamination recognized as a matter of public policy concern and as a matter that needs to be resolved through expenditures for cleanup and mitigation. Given variations in political cultures, rules of law and, perhaps most importantly and 'unscientifically', the strength and concerns of environmental political movements, the same amount of physical contamination in a number of different countries will be recognized as constituting different levels of environmental debt. We suggest here, and will demonstrate later, that the magnitude of borrowing against the environment recognized in any country is less a function of measurable levels of contaminants than of the political processes and values that shape attitudes towards, and responses to, environmental despoliation.

The issue of 'who will pay?' appears at first to be merely a matter of the allocation of economic costs. The potential payees appear at first to be the polluter(s), current title holder(s), or so-called 'deepest pockets', entities with the greatest ability to pay, whether those parties are public or private. However, this three-way distinction, appealing as it is, constitutes an oversimplification.

In a private property-based system, liability for hazards and damages associated with a given property is normally based on land title, since owners are presumed to have benefited from the actions they took that caused others harm. Analogously, product liability law tends to impose a burden on producers for damage done, whether caused by inattention or intent, on the presumption that manufacturers benefit from production and sale of their products. In the case of existing land contamination and its potential health and environmental consequences, however, the appropriate balance of public and private financial responsibility is not clear. At the root of this ambiguity is the inability of the economic system to identify either the specific beneficiaries of such past actions or their share in the benefits produced. In a fashion analogous to the example of a changing capacity to track contamination and its environmental damage in the ground, the 'findings' from economic analysis regarding the allocation of benefits change as better data and more sophisticated techniques become available.

In addition to the broadly shared gains from past pollution, a public policy rationale for extending financial responsibility for cleanup or damage beyond current landowners involves the range of parties that may be responsible for the contamination. Actions taken by companies in the nineteenth century (or even in the post-World War II era), are not logically the exclusive responsibility of current landowners. The perpetrators of the past pollution may no longer be in business or, worse yet, may have inadvertently contaminated a site by their operations on a different parcel of land altogether, since contamination can migrate through the soil.

Finally, no country is wholly isolated at the end of the twentieth century, so both the perpetrators and beneficiaries from contamination may lie beyond the jurisdiction of the national state. The higher the material standard of living of a country, the more likely it is to have at some prior time, if not at the current moment, effectively 'exported' contamination to less affluent states. Evidence from the United States clearly shows such 'dumping' on relatively poor and minority areas domestically (Bryant and Mohai (eds) 1992; Bullard (ed.) 1993; Goldsteen 1993). If such practices are politically acceptable domestically within single markets, they certainly will be internationally. They definitely are an issue for supra-national single markets such as the European Union and areas such as the European Economic Area or the North American Free Trade Area. Economic and political colonies have in the past been victims; now the most tempting location to which to export polluting heavy industry (and pollution) may be the former communist countries of Central and Eastern Europe. In sum, there is no evident efficient or equitable formulaic basis for allocation of responsibility across landowners, polluters or beneficiaries of economic growth. Nor for the spatial allocation of the burden for past or future ground contamination.

Public policies for the environmental mitigation and reclamation of contaminated brownfield sites may thus be expected to vary with a range of sociopolitical factors. Traditional attitudes towards community and individualism, legal doctrines and precedents, and approaches to locality and spatial competition all contribute to shaping the procedures and processes under which removal of past contamination is pursued. The immediate experience of known damage or publicity about dangers associated with contamination can affect political discourse and public priorities and policies. Thus both the definition of the extent of land contamination and practices for the allocation of the burden of the

Introduction: The Contaminated Land Problem

costs for its mitigation may be expected to not only vary from country to country but also over time within any one nation-state.

This variation, and the associated rates of cleanup and reuse of contaminated lands in different countries and under different policies, offers a unique opportunity for comparative analysis. Whether we look at Britain, where the industrial revolution was born, Germany, where the modern chemical industry emerged in the inter-war period, the United States, which built the largest industrial economy of the twentieth century, or any other modern economy, all have experienced some land contamination. (The extent to which such pollution resulted from ignorance or wilful despoliation is not germane here.) In recent years all have begun to address this historical legacy and taken some steps towards repair and reclamation. We thus examine different policies and practices, the factors shaping those approaches, and the specific procedures and responses to the questions of both the debt to be discharged and the allocation of responsibility for payment.

The United States passed the first major legislation designed to address contaminated land, the 1980 Comprehensive Environmental Reclamation, Compensation and Liability Act, known variously as CERCLA or Superfund. (This latter label is used ambiguously in writing about the Act. Technically, it refers to the special fund authorized for expenses incurred by the national government in assuring cleanups of certain high-priority sites, but the literature uses the two terms interchangeably. For purposes of clarity, we will conform to this practice; we use the capitalized word 'Fund' to refer to the special fund.) CERCLA was passed to assure that contaminated lands would be identified and cleaned up, relying on imposition of the legal burden for cleanup costs on all parties in the title of land ownership and on the threat of a central government takeover of the cleanup task. This national law was amended in 1986 by the Superfund Amendment and Reauthorization Act, or SARA. Many of the fifty individual US states have, as permitted by the US Constitution, passed their own legislation affecting either the process of determination of contamination or the allocation of cleanup burdens within their boundaries. This single-country case, therefore, presents significant variation amenable to comparative analysis.

A wealth of data and appraisal of US policy is readily available, given a history of debate on appropriate national responses that extends over fifteen years. That literature is, however, narrowly focused and offers history but few lessons, due to its failure to consider the implications of a broad range of policy alternatives. Typical of the analytical myopia, perhaps, is a 1992 report on *Assigning Liability for Superfund Cleanups* that not once, in its discussion, considered any of the evidence available on the experience under different liability regimes outside the United States (Probst and Portney 1992). By contrast, the British government's *Paying for Our Past* consultation document (UK Department of Environment and Welsh Office 1994) explicitly considered US experience, and the British government in general has been seeking to learn from other examples and experiences in other European countries (Connell 1994).

A spate of recent books have emerged addressing CERCLA and its supposed successes and failures, incorporating innovative uses of data on the spatial distribution of funded cleanups, the political power of legislators from different districts, and the social, economic or racial inequality of the spatial distribution of pollution and cleanup (Hird 1994; Barnett 1994; Bullard 1990). These volumes similarly focus on the United States and, more specifically, on the Federal Environmental Protection Agency, virtually to the exclusion of examination of policy variation between the individual fifty US states. A series of introspective assessments and appraisals can contribute to domestic political debates and may even change policies.

A single country, however, tends to share cultural values and norms and is governed by a common set of legal precedents. It cannot offer the range of contextual variation, nor the variety of administrative and regulatory approaches provided by a multiple country comparison. We thus examine the United States relative to a number of countries in Europe, primarily the United Kingdom and Germany, which have their own cultures, legal histories and experiences of reported damage due to contamination. The European contexts, however, are also shaped, to a degree not evident in the US case, by the policies and influence of a supra-national body, the European Union (EU). EU policies will increasingly affect the policies of its member-states in ways analogous to the influence of the US federal government on the individual states. Thus the examination of emerging EU policy and its relationship to the practices of member-states provide a comparative analogue to the federalism of the US.

COMPARATIVE ANALYSIS AND OUR LOGIC OF ARGUMENT

Comparison between the United States and Western Europe is somewhat complex and must be pursued on two levels simultaneously. Parallels do not exist for US federalism in most of the individual nation-states in Europe, given the autonomous power of the fifty individual states; rather the analogue is the relationship of the European Union to the federal government in the US. On the other hand, the potential for conflict over implementation practice and inconsistencies resulting from different legal systems and cultural norms that characterizes EU policy has as its appropriate analogue the problems faced in implementation of the North American Free Trade Agreement (NAFTA). Thus, we will at times have to compare the fifty US states to a number of the EU member-states, but with respect to other factors shaping contaminated land policy, we will be comparing the US directly to those countries. We will also, in other contexts, compare the US as a whole to individual nation-states in Europe.

We begin our study with discussion in this chapter of some available, albeit sorely incomplete, evidence on the extent of the contaminated land problem, that is, records and reports on the 'objective' amount of contaminated land in different countries. These findings do not constitute specification of individual countries' definitions of the pollution debt they need to discharge, but provide some background on the magnitude of the political problem posed by past and ongoing contamination.

Next, we turn to the issue of the parties affected by land contamination. Obviously, those incurring the most immediate impacts are people and institutions in the locales that were the centres of industrial production for the past century and a half. The historical geographic centres of production are the major cities of the advanced countries. But the problem is not simply urban, as we explain with a brief review of nine different 'constituencies' affected by potential property market distortions associated with having to pay for cleanup of polluted sites.

The fourth and final portion of this chapter lays out the types of financial and investment impacts that need to be considered in assessing policy alternatives. Since our concern is with the reclamation, redevelopment and reuse of contaminated properties and the nature and allocation of the pollution debt to be discharged, we focus on the costs that developers may face in making decisions on the use of brownfield sites that may have been contaminated.

Those costs and development impacts are affected by a wide variety of national factors, which we address and enumerate in Chapter 2. We identify and discuss dimensions across which the policy context for contaminated land policy decision-making and cost allocation can vary. The national and supra-national policy examples in Parts II and III exhibit variation across these dimensions and provide the empirical evidence for our trans-Atlantic comparison of policies and approaches to contamination in Part IV.

Part II, consisting of two chapters, is devoted to the United States. Chapter 3 examines the emergence of land contamination as a political issue, and traces the legislation and regulatory experience at the federal (national state) level. The CERCLA taxation authority that funds the Superfund expires at the end of 1995, so the legislation and its impact have been the objects of extensive scrutiny for the past two years. Proposed CERCLA amendments were offered by the Clinton Administration early in 1994 and a variety of other amendment provisions were proposed by legislators and other interested parties. Chapter 4 examines these proposals and a number of individual state policies and practices to assess the political, social and economic factors impelling them.

Part III describes practice, policy and experience in Europe. Chapter 5 sets the European context for consideration of the specifics of different countries' cleanup efforts. We review the economic and political ties binding the different 'Europes' and examine the movements towards a consistent policy through consideration of the debates and findings of the institutions under the jurisdiction of the European Union. In Chapter 6 we present, compare and contrast detailed findings on the United Kingdom and Germany and seek to develop a Europe-wide perspective by referring to the situation in several other EU states, including France.

Part IV begins with a reappraisal of the experiences of the countries we have examined and proceeds to a series of broad policy recommendations that we consider germane to all contexts exhibited by advanced industrial countries. Chapter 7 returns to the policy dimensions specified in Chapter 2 and compares the American and European findings on each, placing specific legal requirements and regulatory practices into their political and social contexts. This comparison leads to findings on the factors shaping policy successes and failures. The findings are then used in Chapter 8, our conclusions,

Introduction: The Contaminated Land Problem

as we derive a set of principles to guide public policy in directions that contribute to the most efficient, effective and expeditious redevelopment, reclamation and reuse of contaminated land.

THE EXTENT OF THE CONTAMINATED LAND PROBLEM

As we have noted, all the industrialized nations of the world need to correct for past depredations that were legal and accepted when they occurred but violate current standards of environmental protection. The number of sites recorded as contaminated can vary with the size of the country, patterns of landownership, and the extent to which sites have been surveyed for pollutants. Moreover, simple identification of chemical contamination says nothing about its extent or the potential associated damage or cleanup costs.

Nonetheless, the extent of the problem cannot be understated. As the indicative data in Table 1.1 makes clear, the numbers of sites and land area involved are impossible to ignore. National responses, however, have not been consistent, and many countries have yet to formulate coherent approaches to their contamination. Even those with long legislative histories of efforts to address contamination seem uncertain about the magnitude of the problem they face.

The absence of any site data for France underscores the variation present in national approaches. Unlike the other countries for which data are enumerated, France has tended to reject broad central government responsibility to date, leaving the brownfield problem to its *départements*, the regional arms of the central state, and to individual local *communes*, or municipalities. Despite France's apparent unwillingness to forge a coordinated national approach to the contaminated land problem, such national responsibility is presumed in the emerging European Union approaches to environmental issues. Thus France is beginning more systematically to inventory its polluted sites, which must comprise a substantial number, given the EU estimates of the land area contaminated, almost 30,000 square miles.

It should be noted that EU measures of the extent of the contaminated land problem have tended, more often than not, to be expressed in terms of *area*, while the US has generally cited the number of contaminated *sites*. While a number of European nations have looked at both sites and land area, this difference in the units of measurement reflects both policy orientations and attitudes. A site-specific measure is consistent with allocation of responsibility for cleanup to owners, that is to individual parties. An area measure is consistent with acceptance of a broader social responsibility for mitigation of past pollution, one that may ignore the specific ownership of individual sites in a contaminated region.

The United Kingdom is actively in the process of formulating a strong national approach to its contamination, yet it has only crude estimates of the extent of its problems. In fact, under the provisions of Section 143 of its Environmental Protection Act 1990, local authorities in the UK were to produce registers of contaminated land. However, in March 1993, the national government withdrew the requirement because of concerns that the registries would permanently reduce the value of lands that would otherwise command premiums for their prime urban locations.

Chronologically, the US was the first country seriously to address the contamination problem. Since before the 1980 passage of CERCLA, the US has attempted to identify polluted sites, at least those with high levels of contamination or those presenting exceptionally high health risks. The extent of contamination in the United States is massive and the methods of recording known or suspected polluted sites vary across the fifty individual states. Thus, despite over a decade of US cleanup efforts, policy specialists disagree by a factor of 50 per cent over the actual numbers of sites involved.

Availability of data for the EU as a whole on a consistent basis poses an even greater problem than the US has faced in collecting information through its fifty states. The one known attempt to collect consistent and comparable contamination data for the EU member-states, which is limited to former coal and steel sites, is discussed in Chapter 6.

Given the uncertainties regarding the sheer number of sites that may or may not be polluted, we cannot specify the extent or magnitude of the contaminated land problem in terms of land area or numbers of sites. Obviously, if we cannot be specific about those factors, we similarly cannot specify the problem in terms of the volume of contaminants left untended in the ground, nor in terms of the magnitudes of different types of contaminants. However, the problem we are addressing is more than purely environmental or ecological: contamination produces economic development problems as well as adverse ecological impacts and threats to human health. The severity of some of these consequences of contamination can be discussed in more detail.

Introduction: The Contaminated Land Problem

The British concern over the possibility of land stigmatization as the result of past pollution is an economic development issue in all industrial countries considered, including the United States. The historical legacy of contamination is, inevitably, experienced in the oldest and most highly developed settlement areas. Thus, riverfront land, old ports, warehouses and industrial sites, potential prime central business districts and urban central lands, may be unusable until brownfield contamination is removed. To the extent that the presence of pollution is recorded, but no protocols or assurances that the sites will be cleaned up provided by the national state, the value of the land in question, and its ability to attract investment capital, will be depressed. The cost allocation issues associated with contaminated lands and cleanup efforts thus include the spatial redistribution of new investment and future economic development.

Table 1.1 The extent of land contamination

Country	No. of Contaminated Sites
Denmark	10,500
France	75,000 square kilometres
Germany	144,000
The Netherlands	110,000
United Kingdom	50,000–100,000
United States	400,000–600,000

Sources
Lines 1,2,4 — Connell 1994.
Line 3 — CEC 1992b.
Line 5 — Warren 1994.
Line 6 — (low) Bartsch et al. 1991.
           (high) Edelstein 1988.

WHO IS HURT BY CONTAMINATION AND CLEANUP COSTS AND DELAYS?

In Western industrialized societies, this question has tended to have a very limited focus: human beings. We do not address here the issues posed by deep ecology (Devall and Sessions 1985), eco-feminism (Plant (ed.) 1989), the 'land ethic' of Arlo Leopold (1949), or other environmental philosophies and arguments that extend beyond the immediate impacts on humans (Zimmerman et al. (eds) 1993; Platt et al. (eds) 1994). Our admittedly limited focus is dictated by the belief structures and values that bring issues to the fore politically and the policy issues they present. Thus, we accept here a frame of reference limited to the *impact on humans* — and even more narrowly on the people living within a single polity — in addressing the issue of harm associated with the undischarged debt of past pollution.

The spatial economic history of the advanced industrial countries suggests an even narrower focus on a subset of humans and their settings. Factories of 10,000 and more employees characterized the advanced production of the first half of the twentieth century. This economic activity, arguably responsible for the mass of the land contamination that now poses a problem, thus had to be located where large numbers of workers were, or could be made, available — in cities. As a result, with the exception of pollution caused by mining, the contaminated land problem has an urban face.

Cities are resources. That is, they provide a setting that permits people to pursue their livelihoods, to engage in property development and other economic activity. An urban area is more than the aggregate of its physical infrastructure and people, it is a *setting*, or in the broader sense, a physical, social, economic and political *environment*. Any damage to the physical environment of a city undermines the ability of its residents and businesses within it to pursue their livelihoods, to exercise what are, in effect, their property rights in their urban environment. Considered in this manner, cities present the problem of a 'commons' (Hardin 1968). Thus, site-specific contamination, to the extent that it diverts investment away from its environs or drives up the costs of capital to nearby businesses, may be said to undermine a 'common pool resource' (CPR), the sociopolitical environment in which it continues to fester.

By CPR, we mean, following Singh (1994:6), 'resources that are used in common by identifiable groups of people, irrespective of whether they do or do not have well-defined property rights in them'. Such resources are 'property' in the broadest sense of the term, that is, a source or potential source of revenues and economic means traditionally or historically available for people to use in pursuit of their economic well-being, which is recognized and protected by the state. If cities were not property in this sense, then the urban regeneration programmes and policies of Western Europe and North America would have far weaker rationales (and political support). Our

Introduction: The Contaminated Land Problem

concept of a city as a CPR is perhaps best illustrated from the perspective of real estate markets. While each plot of land has a unique location and the structures on each may vary, all are valued in the market with reference to neighbourhood characteristics. Residential real estate, in particular, tends to be valued by assessors using 'comparables', other units in the area that have recently been sold, establishing a market price. All properties in the reference area thus share some common pool resources that produce their common value. The importance of legal or political areas as sources of some CPR is perhaps no more evident than in the unique premium placed on real estate in the City of London; similar distinctions are evident in differences between prices for properties on the two sides of any street that constitutes the dividing line between two jurisdictions.

The impact of policies towards contaminated land on the CPR we call the city or urban area, then, needs to be considered as one of the elements of any measure of the 'success' of such policies. Approaches may successfully impose cleanup and liability costs on past contamination generators in accordance with the 'polluter pays' principle, and may finance state shares of cleanup costs in equitable manner. If, however, in so doing, they fail to arrest damage to the economic and investment climate (or environment) in urban areas through slow resolution of contested claims or unclear definitions of legal responsibilities and levels of cleanup required, we would find them wanting.

This argument places the interests of affected communities in a much more central position than has much of the argument over appropriate approaches to contaminated land, especially those evident in the recent appraisals and criticisms of CERCLA in the US. It forces a definition of community that is much broader than those residents immediately affected by health risks and it includes among those economic actors exposed to involuntary financial risks all those whose livelihood depends on the health of the urban area as a CPR. The potential economic losses confronting would-be developers, their financiers and their insurers over the liability claims on polluting businesses, while legitimate, reflect only a fraction of the impact generated by continuing contamination.

Deterioration of the condition of the urban CPR affects different parties in a number of distinct ways. Consider the following groups

1. Landowners and lenders whose assets or collateral consists of potentially polluted properties, since the value of such lands may

fall. This group also includes owners of lands close to contaminated brownfields, since ground contamination can migrate through soils and underground waterways to other properties.

2. Investors and developers, who may incur additional risks in attempting development in untested 'greenfield' locations because of depressed returns to investment in brownfield sites.

3. Urban residents as a whole, who suffer economic losses due to underutilization of urban lands that are avoided by investors and the associated reduced levels of economic activity within the cities, since land is not used for productive activity. They also suffer from sprawl to greenfield areas and increased travel distances that impose costs both in terms of direct travel time and money and in the environmental spillovers on air quality associated with transportation systems, not to mention the direct effects of possible exposure to the contaminants at the polluted sites.

4. All taxpayers in the nation, to the extent that increased tendencies towards sprawl increases the need for infrastructure maintenance and new construction.

5. All parties interested in the preservation of open space and greenbelts in and around conurbations and metropolitan areas, as abandonment of urban land leads to accelerated development pressures on farmland and other exurban properties.

All these groups pay for urban blight. The problem of land contamination and potential brownfield site abandonment thus cannot be dismissed as a purely 'inner city' issue. It must be addressed in terms of its impact on nation-states and economies in their entireties. The policy question of who will pay for the reclamation of the potentially or actually polluted lands that contribute to that blight is thus central. Any useful answer must transcend the simple tripartite division of polluters/owners/deepest pockets (Yount and Meyer 1994b). As we review the approaches employed in a number of different settings, we will consider the incidence of the costs for reclamation and redevelopment across different types of policy constituents (the groups just enumerated) of the various avenues of response to the contaminated lands problem.

COSTS ASSOCIATED WITH CONTAMINATED LAND CLEANUP EFFORTS

Considering the approach to brownfields as a matter of national and local economic development strategy rather than an issue of environmental policy, it is possible to identify the potential costs and benefits associated with different policies. The possibility of contamination affects property investment in readily determinable fashions, potentially adding to investment transaction costs as well as affecting rates of returns on redevelopment projects involving brownfield rather than greenfield sites. The fact that no consistent approach has emerged across industrialized nations facing comparable cleanup costs suggests either that those financial implications have not been carefully examined or that their effect on investments varies in previously unexamined ways.

Until clear and explicit provisions for reclamation requirements and cost allocations to different parties are promulgated as a matter of law, uncertainty abounds for all parties that have had any relationship to a contaminated or potentially polluted property. Legislation alone may not resolve the uncertainties, since they derive from a new set of socioeconomic relationships. Fifteen years after passage of CERCLA in the US, for example, both lenders holding bad loans and public bodies eligible to take title to lands for non-payment of taxes now take steps to avoid contamination-related uncertainties: they either write off their debts altogether rather than take title to potentially contaminated land or engage in foreclosure only after detailed (and expensive) site investigations to determine the presence or absence of pollution.

Similarly, in Europe, there is growing concern about the extent to which the legacy of contaminated land presents problems not only of reclamation but also of liability. Ambiguities about the legal position of developers and their financiers in the different member-states of the EU and uncertainties in land and property markets inhibit development of brownfield land. The Council of Europe sought to address these issues by inviting governments to sign the *Convention on Civil Liability for Damage resulting from activities dangerous to the environment* (Council of Europe (CE) 1993), known as the Lugano Convention. The European Union, attempting to arrive at consensus

among its members in order to legislate on the liability issue, published its *Green Paper on Remedying Environmental Damage* the same year (Commission of the European Communities (CEC) 1993a). The UK rejected both the Lugano Convention and the Green Paper, possibly increasing uncertainty in British and other property markets. The British government subsequently issued a consultation paper soliciting inputs from UK developers and investors as to what legal arrangements it should promulgate (UK Department of Environment (DOE) and Welsh Office 1994), as a result of which new policy proposals have been formulated (UK DOE 1994).

The effects of many countries' efforts to impose a 'polluter pays' standard, a principle adhered to by both the United States and the European Union, may thus have a major impact on investment behaviours. Two distinct types of costs need to be considered. First are the costs of decision-making, that is the so-called 'transaction costs' associated with acquiring the information needed for decisions and with bringing together all the parties that might share responsibility for past pollution. The second are impacts on rates of return for investments in brownfield as compared to greenfield sites. To the extent that pollution mitigation efforts impose new costs on reclamation and redevelopment of previously used land, such rates of return will tend to be reduced. (Cf. Bartsch et al. 1991, Fogleman 1992, Kelley 1991, Lewis 1991, Mundy 1992a, O'Brien 1989, and Strayer 1992 for the US; Dieterich, Dransfeld and Voss 1993 on Dortmund and Düsseldorf in Germany; Acosta and Renard 1993 on French cases; Needham, Koenders and Kruijt 1993 on The Netherlands.)

Transaction costs may, under some conditions, become so high that the project appraisal itself becomes excessively expensive, making redevelopment virtually impossible. Factors contributing to higher transaction costs include:

a. fees charged for the environmental assessments required by lenders when they are asked to finance the reuse of any previously developed land; especially for small sites, these fees can be a substantial portion of loan costs or proceeds;

b. project delays attributable to site assessments and to determinations of the cost of cleanup; and,

c. increased project underwriting costs and additional legal expenses, associated with assuring that legally-required diligence requirements for minimizing liability risk exposures have been met.

These costs impose burdens on all brownfield redevelopment projects, slowing the adaptation of land use to changing economic conditions and increasing the potential for property market failures. Since these project transaction costs are only partially sensitive to the scale of the transaction, such cost burdens may preclude the cleanup and reuse of the small parcels of potentially contaminated land which constitute the most numerous cases of polluted sites.

In addition to imposing transaction costs, environmental legislation and regulations regarding contaminated land may reduce the perceived (and, sometimes, the actual) returns on investment on such real estate development. These effects may render certain types of investment simply uneconomical, especially in cases in which the appropriate or permitted 'highest and best use' is not a high intensity utilization (Yount and Meyer 1994a). Differences in the political ideology and litigiousness of countries can have exceptional impacts on these cost expectations; the US and UK are the nations most likely to find liability concerns depressing project appraisal values. Factors undermining the perceived potential return on investment include the following.

a. Massive uncertainty over the actual costs of property mitigation. Since each site is unique, each cleanup must be tailored to the specific types of wastes present and to how the site's physical characteristics (geology, climate) determine the effects of contaminants on the environment.

b. Equivalent uncertainty — and, in the US, even greater fear — over liability for environmental damage resulting from past uses of the land and contaminants in the soil. This concern is accentuated in contexts in which the liability standards themselves are unclear.

c. Possible property stigmatization, associated with publicity over specific sites having been contaminated or having required environmental mitigation. Fears that sites may be labelled as 'polluted' and suffer reduced market value even after approved cleanups led the UK, for example, to reject the idea of a register of contaminated land.

d. Reputational losses, or stigmatization of developers and their financiers found to have overlooked or inadequately mitigated contamination, or who have had projects fail due to unanticipated cleanup costs. The returns on any one project may not be sufficient to risk the potential loss of institutional reputation, since such a loss could affect future access to capital.

e. Lender and other investor-imposed or legally mandated deed restrictions, restrictive covenants, or pollution-specific land use or zoning restrictions that limit sites' future uses. (Post-cleanup monitoring requirements can continue for several decades and may generate a need for recorded easements and other deed restrictions that limit the future development and use of an affected property.)

f. Potential post-cleanup costs associated with prospective (future) liability for past pollution. If mitigation approvals are subject to being reassessed in the future, such costs can arise from conflicting and changing standards for mitigation over time as governments raise requirements and regulations in response to improved contamination detection and mitigation technologies.

g. Finally, in an uncertain regulatory climate, investors fear the risk that an administrative agency enforcement action may result in a public taking of their assets or collateral. They thus would tend to further discount the projected cash flow of an investment.

In combination, these cost expectations and fears lead investors to reduce their valuations of expected investment returns. As a result, many projects that could be both environmentally sound and economically viable may not be undertaken. Both individual municipalities and the economy as a whole would thus be deprived of possible welfare gains due to the uncertainty and potential liability imposed by contamination cleanup requirements.

What we have referred to as a debt on past pollution must be paid. This is a principle on which the European Union and the United States are agreed. However, the costs of raising the requisite funds for such repayment can vary substantially depending on legal requirements and the procedures for cost-sharing. This brief enumeration of different investment cost considerations identifies factors needed for assessment of current policy practices; we return to them in Chapter 7, as we compare the US and EU experiences.

Notwithstanding the importance of investor appraisal of project alternatives, we must remember that their 'cost' is not an absolute. The *public* imposition of an obligation on one party by laws or regulations that constitutes a financial burden does not preclude the *private* reallocation of those expenses through higher prices, liability insurance claims or other processes. The institutional and social contexts in which public policies are generated and implemented thus affect the ultimate pattern of cost-sharing as well as the total costs involved. We turn next in Chapter 2 to an examination of how those contexts can vary for public policies towards contaminated land.

2. Dimensions of National Contaminated Land Policy Contexts

The emergence of concern for the environment has taken a variety of political paths and forms in different countries and has had very divergent histories. Cultural values, legal traditions — especially the form of property rights and the forms of governance — largely the interactions between the national state and more local governmental entities, the reliance on market processes, and myriad other national practices all combine to shape environmental control and preservation regulations. In this chapter, then, we consider the dimensions along which these contextual factors can vary. Each could, and did, contribute to the practices in control and reclamation of contaminated land that have emerged in recent decades.

Yet another factor is the sheer volume of land. With its vast size and underutilized open spaces, the US could, to an extent not possible in the much more densely populated countries of Europe, be complacent about its supply of 'spare' land. Profligacy may be bred by surplus supply; certainly, from an economics perspective, it is scarcity that drives the development of allocation schemes and requirements.

We set out in this chapter a framework for relating contaminated land policy approaches to the broader socioeconomic and legal context in which it is implemented, considering a number of different dimensions. The policy environment is complex, and we make no claim to having full data on all elements for the settings we compare. However, their use can suggest ways in which approaches to addressing contaminated land may be modified to produce improved outcomes.

These dimensions can be grouped and subdivided along the lines of Figure 2.1. This chapter follows the organizational logic of the figure, beginning with past experience of environmental problems, then turning to the formal legal context in which issues are addressed, the forms and structures of land-use control systems in general, and the national approaches to legal liability, before concluding with a look at experiences with current regulatory approaches and the political strength of environmental movements as the final factors shaping the debate about appropriate means of addressing the amelioration of past land pollution.

Our discussion here initially considers a number of dimensions in the context of private ownership of land, except where we expressly consider various arms of the state as landowners. Much of our discussion of 'private' responsibilities below, therefore, applies to public sector landowners or users who do not draw revenues from the general budget of their respective national states. Their financial condition and problems when confronted with costs from past contamination are akin to those of private firms.

PAST ENVIRONMENTAL EXPERIENCES: ACCIDENTS, EXPOSURES AND 'DISASTERS'

Past experience clearly sets the environmental agenda to the extent that it affects public opinion and political imperatives. The influence of any given events or reported exposures on public policies, however, is not strictly a matter of their severity or the threats they pose. Public perception is not necessarily correlated closely with either the technological hazards present or the costs of their mitigation. General publicity, local media coverage and factors such as the number and reported severity of pollution-generated problems, nature of the pollution, the organizations responsible for past contamination, community capacity to produce cleanups and speed of public sector response all help shape the perception of the problem of past land contamination.

Local responses and concerns vary within countries. Areas with longstanding environmental exposures or risks may become inured to the dangers. Areas with apparently safe operations of risky technologies may not express concern about such activities while other locations may resist their incursion.

While fears may be raised by instances of contamination, public sector actions do not necessarily follow. The immediacy of the threat and the perceived cost of response affect the extent to which such an issue may rise to the top of the political agenda. The discoveries of dioxin pollution at Love Canal, New York, and Times Beach, Missouri, in the United States, and the dioxin spill at Seveso, Italy, and the Rhine

Figure 2.1 Dimensions of the contaminated land policy context

PAST ENVIRONMENTAL EXPERIENCE: ACCIDENTS, EXPOSURES AND 'DISASTERS'

THE STATE POLITICO-LEGAL CONTEXT
  Mandates from Central Jurisdictions
  Reliance on the State and Markets
  'Local' Options and Flexibility
  Spatial Policy Approaches
  Standards for Cleanup

THE LOCUS OF CONTROL OVER THE USE OF LAND
  Exclusivity of Property Rights
  Zoning Approaches
    • Legal Immutability
    • Zoning Logic
  Land-Use Controls
  State Powers to Take Land
  Preservation and Development Limits
    • Architectural Constraints
    • Preservation of Open Space

APPROACHES TO LIABILITY FOR DAMAGES AND CLEANUP
  Environmental Damage Liability
  Cleanup Liability
  Liability Allocation
  Treatment of New Landowners
    • Required Disclosure
    • New Purchasers' Liability Shares
  Treatment of Future Liability
    • Public–Private Sharing
    • Private Sector Cost-Sharing
  Public Landownership
    • Total Public Landownership
    • Sub-National Public Landownership

Figure 2.1 Continued

ACCESS TO CAPITAL AND TO LIABILITY INSURANCE
  Private Sector Lending Practices
    • Centralization of Financial Institutions
    • Financial Institution Risk Aversion
    • Recent Experience with Risky Loans
    • The Availability of Lending Insurance
  General Liability Insurance Practices and Experiences
  Experience of Spatial Displacement and Capital Flight
    • Actual Losses to External Locales
    • The Extent of Land Dereliction and Abandonment

CURRENT POLICY AND POLITICAL PRESSURES
  Experience under Current and Past Cleanup Efforts
  The Strength of Environmental Movements and Green Parties

---

fire in Europe, for example, played different roles in generating public sector policies in different nations, as the next two parts of this book illustrate.

Whatever the actual experience on the ground, responses are shaped by a variety of national and local factors, as we have noted. We turn next to the politico-legal context of policy-making, dwelling on the extent of centralization or local specificity of policies and standards for environmental performance.

THE STATE POLITICO-LEGAL CONTEXT

We begin our discussion with a review of the character of possible mandates from the central jurisdictions considered here (EU and US) to more local jurisdictions (EU member-states, the US states, and local and regional governments). Next we focus on roles of the local state, before considering the extent to which local flexibility is available, the policy spatial focus and the rigidity of cleanup standards themselves. Each of these factors shapes how knowledge of environmental problems and publicity about adverse events translates into specific policy responses.

Mandates from central jurisdictions vary in their rigidity and the policy space available for flexible local response. Such mandates have arisen as a contentious issue in the United States. The fifty US states, over recent decades, have been assigned ever more responsibilities and obligations by the national state. The rationale for these mandates is that actions taken in any one state affect other political jurisdictions and thus a common standard is necessary. A further argument for such mandates in the US context has often been the need to keep states from competing for economic development through relieving businesses of the cost burden of the maintenance of appropriate environmental standards.

This issue has risen in significance in the European context with increasing integration of policies and the growing power of the European Union to impose standards on its member-states, as well as the creation of the Single European Market, which increases the possibility of production relocation to weaker regulatory settings. The analogue to the relationship between the nation and the states in the US federal system, therefore, is the relationship of the European Union to its member-states. The new North American Free Trade Agreement (NAFTA) potentially parallels the European situation, providing argument for the US to share common standards with Mexico and Canada, dictated by the environmental agreement that accompanied NAFTA; unlike the EU case, however, NAFTA is inferior, not superior, to national laws and regulations.

Central standards encompass provision of procedural guarantees and assurances of particular *environmental control processes*, imposition of specific standards for *environmental outcomes*, or both. The processes of contaminated land cleanup are shaped by the relative emphasis on procedures and outcomes, which, in turn, is influenced by both the legal rights inherent in landownership and those to a clean environment.

Reliance on the state and markets for pollution mitigation and redevelopment between them varies. The United States is arguably the most market oriented of the advanced capitalist countries. It is, for example, virtually alone in imposing no national regulation on the spatial allocation of investment through regional policies or other controls.

The United Kingdom is arguably the most market oriented of the European countries and has progressed the furthest in the dismantling and privatization of state-owned enterprises. Belgium is not far behind, at least with respect to land and property markets. Most of the other

EU member-states are far more corporatist in character, with the state accepting a higher level of responsibility for economic activity, production and well-being. France, for example, continues to exhibit extensive public ownership of heavy industry and control of banking institutions. The supply of industrial land in The Netherlands is almost completely under the control of public authorities. Germany, even before reunification, exhibited a higher level of state acceptance of liabilities associated with environmental damage; since the absorption of the former German Democratic Republic, the legacy of communist central planning has increased the recognition of state responsibility, as our example of Bitterfeld in Chapter 6 will illustrate.

The majority of European nations, moreover, are unitary states, with centralized power at the national level. The main exceptions exhibiting some federalism akin to that of the United States are Germany, Belgium, Switzerland and, to a large extent, Spain. We ignore here the difference between such federal systems and the unitary states, since the accepted responsibilities of government carry more weight in shaping their environmental roles than do their specific legal structures.

The traditions governing state responsibility for social welfare, broadly conceived, clearly shape practices regarding state acceptance of responsibility for environmental protection. Thus, the critical distinction is the extent to which individual welfare is seen as shaped by systemic forces as distinct from individual choices. Individualism will tend to generate an approach to contamination that minimizes state roles, including enforcement as well as cost-sharing for liability and cleanup costs. Full state provision, by contrast, would tend to generate strict regulatory regimes and provision for liability and cleanup costs through taxation and public sector spending. The former may engender uncertainty over costs that derive from private sector actions (including the 'anarchy of the market') and court decisions on litigation. Full state provision for mitigation on the other hand, while it would reduce uncertainties, would also eliminate financial incentives for contamination minimization by current land users. Neither extreme is thus conducive to responsible land use and reclamation of contamination, and the appropriate balance will depend on a mix of other factors.

'Local' options and flexibility involve the power of sub-national units in the US, or the power of such units and the member-states themselves to disregard or modify the EU standards in Europe. At one extreme, the central standards govern, with no flexibility at the local or regional level. The alternative would vest power in the local units

to ignore or modify central standards to conform to those of a local constituency. Thus, local jurisdictions might vary cleanliness standards for different intended land uses even if a common standard were mandated by the central body — or might either waive procedural requirements, or impose additional ones, depending on local perceived needs and imperatives.

There are actually three alternatives in the possible provision of flexibility to the local state. First, none may be permitted, as the central standards strictly govern both procedures and outcomes. Second, in those instances in which power *is* vested in the local state to modify central mandates, it may be limited to the provision of power to the local state to *strengthen* central standards, but not to weaken them. The third possibility is to provide full power to the local state to both raise or lower standards for processes and outcomes, leaving the central body in a strictly advisory role. In practice, then, the continuum is from governing to advisory standards. The flexibility provided the local state may be different for procedures than for outcomes, may vary for different environmental concerns, or may vary for different aspects of environmental regulations.

Spatial policy approaches can vary depending on legislation and on common practice or tradition. Whole districts may be polluted and may share common sources of contamination, but they may be addressed either as regions or on a site-specific basis. Nations with strong traditions of regional economic policy and central state control over or direction of development may be inclined to adopt district-wide approaches towards mitigation of past contamination. Those that have relied on market forces would normally tend to continue their practices and address past contamination site by site, relying on individual owner or developer initiative. Only in extreme circumstances, where a number of sites all contribute to a common polluted pool of groundwater or otherwise cannot be considered separately due to physical factors, might such societies deem it appropriate to address an entire area simultaneously.

The appropriateness or effectiveness of any spatial policy tradition in addressing past land contamination may depend more on the nature of the pollution and past land uses than on other cultural or political factors. The regional or district spatial policy approach facilitates state action, but undermines individual initiative where profit potential exists. Where pollution is widespread and domestic real estate investment alternatives plentiful, contaminated regions may be effectively abandoned by private capital, so state-led recovery efforts may be essential. (This condition appears to be present in the United States, but the state has not accepted such a lead role, as we will discover in Chapter 4.) Countries in which land use has tended to be intensive — typically smaller nation-states such as those in Europe — may be more inclined to take district-wide approaches.

Standards for cleanup, that is, the maximum permissible remaining levels of contamination, might be associated with the extent of state regulation. However, even in the extreme case of full state responsibility for cleanup, there exists the possibility of differentiated approaches. On the one hand, a single standard may be imposed for mitigation of pollution. At the extreme, this standard might require restoration of all contaminated land to 'background' or pristine conditions that would be presumed to have existed prior to human use of the land for production. Such an approach would assure minimization of any future threats or damage to either ecosystems or human health, but would, obviously, be the most expensive possible response to the presence of undesirable pollution.

The alternative to a single requirement, whether strict or lax, is a series of flexible standards. Cleanup requirements might vary with the type of contaminant or may be linked to the intended future uses of the land found to be polluted or of sites adjacent to the one known to be contaminated. Land intended for residential use, schools, childcare centres and the like might require more removal of potential toxics than property that was set aside for industrial use, in which case paving and the construction of foundations for buildings might provide adequate protection from health threats.

Flexibility holds the promise of lower cleanup costs relative to a fixed standard. However, this promise is realized only in the short run, when the intended land use is fixed and known, and while pollutant-type-sensitive standards are not affected by advancing knowledge about risks (or by improved detection capacities). Over the longer term, new costs may arise from this approach: first, the need to maintain records about the extent of reclamation and, second, further mitigation of pollutants in the event of future land-use changes. The imperatives of local economic development concerns combine with a concern for 'environmental justice' to suggest that this flexible form of timed mitigation may not be politically possible due to the uneven spatial distribution of contaminants.

THE LOCUS OF CONTROL OVER THE USE OF LAND

Land-use law is a relatively recent development in formal legislation. However, litigation and criminal law findings have produced legal precedents for centuries, and cultural norms of privacy and the relative importance of individual and community rights have regulated land-use practices as effectively as any legislation. In some contexts, state powers have been derived from the historical powers of regal sovereigns, in others they have been constrained by written or implicit constitutional or other constraints. In all countries, however, centuries-old traditional land-use control values and twentieth-century zoning and land-use control laws and legal precedents combine to further shape the context in which cleanup of contamination can be ordered and implemented.

Exclusivity of property rights is a matter of both law and custom, and thus varies across countries. It can be divided into two elements, (a) landowner powers to do what they want with their land, and (b) non-landowner rights to access and use privately owned properties. The former is regulated by zoning practices, land-use regulation and public 'takings' powers, which are addressed as distinct policy dimensions below. We limit our discussion here to the issue of non-owner access and use. The concept of 'trespass' varies greatly, with some countries and US states permitting landowners to bar people totally from their lands, while others allow virtually unlimited access, provided crops and facilities are not disturbed or damaged. The greater the general public access and use of lands that is permitted, the more difficult it becomes to tailor cleanup standards for intended uses. While a less stringent cleanliness standard might be appropriate to land used for industrial purposes, public health concerns may dictate imposition of higher standards to protect incidental users, if any, from inappropriate exposure to potential toxics.

Factors that we do not associate with legal issues in land-use control may well play a role in producing the norms of freedom to access land. For example, the US is the strictest nation in limiting public access to private land — but it has the most land and the greatest expanse of publicly owned park lands and preserves. The fact that coastal states in the US tend to constrain landowners' privacy rights and permit others' use of their properties to access the public seaside suggests that scarcity, more prevalent in Europe and characteristic of beaches in the US, may induce the state to permit more access to private lands. (Of course, the free availability of guns in the US may explain the greater tendency of landowners to attempt to enforce their rights — but gun control is not our subject here.)

Zoning approaches vary along two dimensions. First, there is the legal standing of zoning or zones, development plans and building plans. Second, is the logic employed in zone designation and the imposition of restrictions.

Legal immutability relates to the extent to which plans, once promulgated, are legally binding. Only if they *are* binding can they be considered part of a zone planning system, and can provide certainty regarding future land uses. Most of continental Europe accepts a Roman law system with legally binding plans, rigidly so in law in many instances. The UK and Ireland both have discretionary planning systems in which land-use plans are indicative rather than legally binding: while specific plans require prior (local) government approval, the zoning of any given tract is an indication of state preference, not a formal requirement. The United States has even less legal rigidity, with zoning variances available for any plot of land, regardless of its formal designation, and generally minimal expectation of prior state approval of building plans if they conform to the requirements for the approved uses.

Zoning logic is reflected in implementing legislation and regulations. Two basic logical alternatives exist. First, there is the practice of *hierarchical* zoning, common in the US, in which limits are placed only on the most intensive use to which sites can be put (which is, effectively, the extent of environmental disruption permitted). The hierarchy flows from least intensive, single family residential dwellings on large lots, through successively higher density housing to commercial and office uses and on to light and then heavy industrial activities. Under such provisions, landowners could place single family homes on industrially zoned properties.

Preservation of sites for particular uses, by contrast, is the European norm under *exclusionary* zoning, the assignment of a particular use to a given site. In some instances, these designations can be extremely narrow, such as commercial usage limited to professional offices rather than retail sales. This practice is much more amenable to the establishment of the variability in cleanup standards previously discussed, since the intended use which may determine the mitigation standard is much more obviously constrained.

Exclusionary zoning, however, limits the flexibility of landowners to earn profits or maximize economic value from their properties. As a result, international development capital may tend to flow towards countries with hierarchically zoned land, since the long-term options for profit maximization are greater. However, the potential for private speculative landholding in anticipation of future gains is simultaneously greater in those settings; since speculative behaviour involves holding land relatively idle (or in the form of open parking lots or in other underdeveloped states), the hierarchical system, although it may attract more capital, may not have it available to accelerate current redevelopment of previously used lands.

Land-use controls, that is, public regulation of landowner actions undertaken within the constraints of existing zoning controls, also vary. There may be a 'presumptive permission to act' under the zoning and building codes prevailing when a property is purchased. Such an approach may even require compensation of property owners if new constraints on their uses of the land are promulgated for environmental protection purposes. The alternative practice is to presume the right of the state to regulate all land use, and to require the 'active permission to act' from the local state.

The contrast between these two approaches underscores, in effect, the values a society places on individual relative to community rights. A 'presumptive permission to act' places primacy in the individual, while the 'active permission to act' principle recognizes that community interest in the uses of land supersede those of the individual landowner. This difference in the relative importance granted to individual and collective interests is reflected further in two other aspects of the state's rights to control the use of land, which we examine next.

State powers to take land always exist in some measure. What varies, however, is the power of the state to take land from those who would rather not sell, or to establish a 'fair price' for the land it acquires. On the one hand, the state may be vested with 'compulsory purchase powers' (the British term) or expropriation rights (Euro-English) for lands to be used for *any state purposes*, including redevelopment and resale for new private uses. On the other hand, the state's use of its eminent domain powers (US-English) to acquire land may be limited to the right to acquire property to these sites needed for the construction of *specific public facilities*. This difference may not appear major, but in practice it has had substantial importance and the implications of the distinction for state power to acquire property rights for environmental purposes are substantial.

At one extreme of constrained public land acquisition powers, the state may not even be enabled to acquire a parcel of contaminated property for the purposes of cleaning it unless it is to be *employed afterwards for a state activity*. Even if the state could acquire the polluted property itself, however, it may not be able to acquire threatened but not yet polluted adjacent parcels. The presence of residences on those adjacent properties, for example, might then impose a cleanup standard on the polluted site that exceeds that which would otherwise be appropriate under cleanliness standards tailored to other intended adjacent land uses, made possible by relocation of the residents. The more constrained the powers of the state to take land for broad public purposes, the more difficulties will lie in the way of state-led reclamation, regeneration and reuse of contaminated lands.

Preservation and development limits are a relatively recent addition to the array of land-use controls, and constitute a separate class of formal constraints. Two major types of limits on new land uses can be distinguished, architectural constraints and the set-aside of land as open spaces or parks.

Architectural constraints include preservation of buildings as well as limits on the characteristics of new construction. All countries now list and provide legal protection to historic buildings, but the countries of Europe appear to have engaged in architectural preservation informally through cultural norms more than has the United States. Obviously, the larger the stock of existing structures covered by such restrictions, the greater the problem of coping with contamination that may have been caused by prior occupants or uses. In some instances, it may be physically impossible to treat some forms of past pollution while leaving buildings standing. At a minimum, the form of reclamation or mitigation of pollution, that is the choice of cleanup technologies available, may be constrained by the need to retain structures *in situ*.

Preservation of open space by definition precludes the development of certain properties. However, the impact of such set-asides on the costs of and potential for cleanup of contaminated land may depend more on when the properties were designated as reserves than on any aspects of current policies towards pollution mitigation. Urban parks established prior to the late nineteenth century, for example, are not likely to have been seriously contaminated by any of the uses to which they were previously put. Greenbelts in or near urbanized areas are similarly not likely to present a problem in contaminated land cleanup, since they comprise greenfield properties.

However, such land set-aside practices in aggregate can affect the extent of contamination cleanup through their effects on land availability. Wide greenbelts may impede suburbanization by creating a spatial barrier and extending commuting lines. Major central city parks reduce the supply of central city land available for development. The financial returns to investment in reclamation and redevelopment of polluted properties, other things being equal, will be greater in a city with land scarcity than one with abundant property available for development. The reservation of open space or greenbelts inevitably tightens the land market and thus creates a greater incentive for reuse of brownfield properties than would otherwise exist in a city or urban area by limiting the availability of investment opportunities on non-polluted land.

APPROACHES TO LIABILITY FOR DAMAGES AND CLEANUPS

We turn here to practices regarding assignment of liability for past pollution. The allocation of liability across the range of parties who might be considered potentially responsible for the creation and impacts of contamination will inevitably have roots in prior cultural norms and legal practices. Whatever the past practices, new allocation choices exist for the introduction of liabilities for land contamination. The approaches taken may affect the impact of efforts to clean up past pollution on the rates of reuse of previously utilized sites and on urban regeneration overall.

We begin by looking at liability for damage done and for the costs of cleanup and then consider the issue of allocation of those costs across parties sharing responsibility for them. Next we consider the liability status of new purchasers of actually or potentially contaminated sites. We then turn to the issue of the state as a landowner before considering the uncertainties involved in future liabilities that may arise from new discoveries about pollution and its effects, rather than from new acts of contamination.

Environmental damage liability allocation involves assignment of responsibility for the harm caused by contamination — whether to ecosystems or to human health — across potentially responsible parties. The liability could be imposed simply on the current owners and/or users of land found to be contaminated. The extreme alternative is the imposition of shared liability on all current and past owners and users of a site.

The rationale for the former practice includes its simplicity: the current owners and users are easily identified, clearly exist and can be presumed to have resources that could be tapped to pay for damage. This practice, however, imposes a burden on current occupants of a piece of property for practices in which they may not have engaged: prior users of the land may have produced the contamination that is now recognized as a problem.

Shared liability allocates responsibility across all users of a property who may have contributed to the current levels of contamination and is thus more equitable in principle. However, prior users may be difficult to identify and, even if known, may no longer be in business or may be insolvent. Moreover, inclusion of *all* prior users may involve many whose contribution to pollution is minimal and thus may produce a huge administrative cost burden, one that could swamp the actual liability costs.

Cleanup liability, the responsibility for mitigation and/or removal of the contamination found on a site or polluted area, can vary along similar dimensions. The relative advantages of positions from burdening only current users to sharing responsibility across all past users are analogous to those for environmental liability. In many instances, however, the actual damage done to date may be insignificant relative to the costs of mitigation and avoidance of potential future damage. Thus the assignment of liability for cleanup may be more significant than the distribution of liability for environmental damage done. The two different types of financial responsibility tend to be allocated in parallel, but need not necessarily take the same form.

Liability allocation practices can themselves vary, so the practice of including multiple users of a property does not fully determine the costs of such allocation or the availability of resources for payment of damage claims or cleanup. On the one hand, each current and past user may be held responsible for only the known damage it itself generated. Alternatively, all liable parties may be held to strict joint and several responsibility for costs, that is, all polluting parties may share the liability and any one could be assigned responsibility for the total cost if the others cannot pay their shares.

In the former instance, equity considerations would require very detailed determination of the percentage of the damage done — and the cleanup costs engendered — by each past or current use. Since damage may not be proportional to the costs of mitigation for different uses, this practice may require different allocation schemes for each of a potentially large number of past acts of pollution. One firm may have contaminated land with two or more different chemicals, one of which may be easy to clean and the other(s) expensive to mitigate. Many firms may have dumped the same compounds, so the proportion of the chemical deposited by each would, in principle, have to be determined.

A formulaic allocation approach may be logically impossible, given the known chemistry of contamination and available cleanup technologies. First, tons, or volume, of contaminants, whatever their toxicity, do not necessarily translate proportionally to cleanup costs. Second, and far more problematical, is the possibility that a chemical reaction between different benign compounds left in the soil by two or more prior land users may result in the formation of new substances that *are* toxic. The inability to resolve ambiguities in assigning explicit liability for actions has led to an alternative that may be equally difficult to apply in practice.

The imposition of strict joint and several liability, under which all prior owners and users are responsible for all damage done, regardless of blame, does not avoid this cost allocation problem. Civil law may leave responsible parties free to sue each other under pre-existing fault-based liability statutes. In such instances, the so-called 'deep pockets' — the most financially capable parties sharing the liability — may become responsible for others' shares of the cost. The risk of incurring joint and several liability may impede cleanup, since those economic actors that have extensive resources (developers and their financiers) may actively avoid any new involvements in potentially contaminated land, especially if liability extends to include new landowners.

Treatment of new landowners, that is, of new purchasers of land that may have had prior contamination, reflects both a society's attitudes towards the working of the market and its consideration of the state's obligations to private citizens' economic welfare. Such treatment incorporates two elements that may, in fact, be considered to be distinct dimensions, the disclosure required of current owners offering property for sale, and the extent to which new purchasers share in potential liabilities.

Required disclosure can vary from *caveat emptor*, the 'buyer beware' principle, to requirements that sellers fully assess both the presence and extent of past pollution as well as the costs of cleanups and damage liability potentials as a condition for placing their land on the market. The absence of disclosure imposes minimal costs on sellers and may maximize the rate at which land is offered for sale. However, the risks attendant on such a practice may discourage buyers from investing in *any* previously used land, for fear that it may be contaminated. Full disclosure, however, imposes massive transaction costs on sellers, since a full determination of not merely the probability of pollution but of its extent and the costs of mitigation can be both time consuming and expensive in terms of funds required. Therefore, at any given level of public sector participation in the costs of pollution-level determination and contamination remediation, the selection of a level of required disclosure will either retard brownfield supply or depress demand for previously developed land. The optimal mix on disclosure will vary across settings and local real estate market conditions.

New purchasers' liability shares clearly affect the impact that any given disclosure regime has on real estate markets. If buyers are completely absolved of any responsibility for past pollution, then, regardless of the disclosure provided, the only cost they may incur is that associated with time delays in redevelopment and reuse caused by the need to reclaim and mitigate contamination. On the other hand, if purchasers share fully in past and current liabilities then the extent of disclosure will be the determinant of the demand for brownfield sites. However, there is a further consideration with respect to *future* liabilities that may arise from new discoveries of either the toxicity of existing pollution or from discovery of previously undetectable contamination. Their perceived liability exposures, however, will depend more on the treatment of prospective liability than any other factor.

Treatment of future liability arises as an issue since what appears at present to be an adequate mitigation given current technological and analytical capacities may turn out to be insufficient, given additional knowledge of environmental or health risks. Damage may result from exposures that would be permitted under current standards, so further corrective action and compensation for damage attributable to the pollution may be appropriate in the future. At issue is the distribution of responsibility for such newly discovered problems, initially between the public and private sectors, and only then among a number of private actors.

Public–private sharing can range from full state responsibility for all costs associated with newly discovered dangers to full private sector and landowner/user responsibility. Both extremes present problems for policy and incentives. If the public sector bears all the costs, private landholders are induced to (a) do the minimum mitigation that satisfies current standards regardless of the risks that additional cleanup may be needed in the future, and (b) maximize current profitability from the reuse of land regardless of the effect of the choice of new land uses on the probability of a need for further mitigation of pollution in the future. These incentives clearly do not minimize the *social* costs of past pollution or of its mitigation as new dangers may be discovered. They promote socially irresponsible actions to maximize *private* profits under conditions that contribute to potentially great divergence between public and private rates of return on redevelopment alternatives.

Full private responsibility for the potential future costs, however, introduces a high level of uncertainty into any investment appraisal. This uncertainty would not be present for greenfield sites, so it would tend to divert capital away from brownfield sites, other things being equal. Placing the full cost burden on the landholders, therefore, may be counterproductive from the perspective of the public treasury: future mitigation costs on some sites may be avoided by placing the burden on landowners, but other contaminated sites may be abandoned by the private sector with the full current mitigation cost burden shifting to the public sector if action is needed to protect human health and the environment.

Private sector cost-sharing for future contingent liabilities can vary from full sharing across all parties liable for current required mitigations and damage costs to assignment of cost to only current (and potential future) landowners and users. This cost allocation decision is, in part, dependent upon the accepted principles for sharing costs for the cleanups currently required. Sharing future costs across all parties who might have contributed to current known pollution may appear equitable and appropriate. The practice would impose extensive new transaction costs, including the continued tracking of all identified responsible parties. The uncertain possible future costs associated with technological advances in detecting and analysing the effects of past pollution will tend to undermine the viability of *any* firm burdened with such liabilities. The attendant instability and possible reduced rates of aggregate national economic growth could, in turn, undermine the capacity of the economy as a whole to act on the dangers presented by past contamination.

The imposition of all future cleanup and damage charges on current and future users of a property could have similar dampening effects on private sector development activities on contaminated land. In this instance, the cost burdens would be concentrated in those entities undertaking to reuse brownfields. They would never be clear of uncertain future charges and thus could never be certain that a given investment in mitigation and redevelopment would prove profitable. Obviously, if this condition obtained, then no private sector capital would ever flow to the reuse of such sites: profit expectation must exist for investment to occur.

Future liabilities are unknown and unpredictable because they may be shaped by technological advances that cannot be foreseen. If redevelopment investments depend on the types of cost calculations described in Chapter 1, it follows that prospective costs *have to be limited* if economic growth is not to be sacrificed — or if the land currently known to be contaminated is not to be abandoned totally by the private sector. Some public sector sharing in future costs will be required to place a maximum on private cost exposures. The extent of the share required, however, cannot be determined 'logically', as it will vary with the position of the society, economy and legal system on many of the other dimensions discussed here.

Public landownership is common in all countries but its extent varies substantially and the owning public entities take significantly different forms. Any government will generally own at least some portion of the properties housing its offices, bureaus and other facilities. However, we need to make some distinctions here with respect to what we mean by 'government'. Differences in the sources of the revenues that finance public activities exist between levels of governments, that is the nation, intermediate states as may exist in federal systems, and local authorities such as cities and counties. But they also exist between entities executing broad-ranging functions and those in existence for the delivery of specific services. Finally, variations in revenue sources distinguish purely public entities from the mix of private–public organizations that has grown rapidly in the nominally capitalist industrial states in recent decades. What constitutes the public sector appears to vary substantially between different nation-states. In order to better specify the significance of the dimension of public landholding and its implications for the effective incidence of contaminated land cleanup cost allocation policies, we need to examine the differences in some detail here.

In the case of the United States, we find a complex mix of what may be called general government units and single- or special-purpose governments. The former are engaged in a broad array of activities, financed predominantly by broad-based taxes on all people, property owners or businesses within their compass, with some possible earmarked revenues or special fees. The latter exist to execute specific functions or provide distinct services, typically financed by special taxes or, sometimes exclusively, by user fees. Examples of single- or special-purpose units in the US include school districts, sewer and/or water districts, waste disposal authorities, economic development authorities, housing authorities, transportation authorities, parking authorities and even cemetery districts, all of which hold some property required for the delivery of the services they provide.

In most European contexts this distinction between general- and single- or special-purpose jurisdictions is not common, or is not explicitly acknowledged as such. (Any local authority could, in some manner, be considered a special-purpose entity, in so far as its mandated functions are specified and discretionary powers limited by the national state.) Discrete authorities with single or a limited number of purposes tend not to exist except as subordinate agencies of the 'general-purpose' state, reliant upon it for revenues.

One important exception is Britain's Urban Development Corporations (UDCs), which have been vested by the national state with land-use planning powers within their designated areas that are removed from the local authorities within which the UDC territories lie. The UDCs can pursue profits from real estate dealings on the properties in their districts that augment their (national state) revenue streams. There is thus an extremely obvious potential for conflicts of interest between the UDC land-use control and economic development functions with respect to contaminated lands responses. These potential conflicts arise similarly for local general government units and even private–public partnerships, so long as they combine control and development efforts.

Land banking by an arm of the local or regional state, which is an economic development practice present on both sides of the Atlantic, exhibits just such a combination, but not as overtly as UDCs. Whatever the precise legal form of the land-banking entity, its acquisition of land title may expose it to liabilities for cleanup and thus the issue of cost shares, but its acquisitions are driven by a mandate to support development. Similarly, public provision of housing, even for middle-class households, has been widespread in much of Europe. The providing entities range from departments of local government units to publicly sanctioned and supported cooperative housing authorities and other public–private mixtures. Rents provide the major revenues, sometimes supplemented by general local government revenue appropriations (in the case of public agencies). To the extent that economic development agencies or housing providers share responsibility for past pollution on lands they hold, they pose questions analogous to those associated with private landowners: what share of their costs should be incorporated into their fees and what proportion borne by the general revenues of the 'parent' public sector unit?

Other public and quasi-public landowners abound. While they are being privatized in many countries of Europe at present, a wide array of public utilities (electricity, water and sewer, telecommunications, railroads, urban public transportation systems) have, historically, been delivered by public corporations that received their revenues primarily from user fees. These fees have generally been augmented to varying degrees by state subsidies designed to hold down direct user charges for at least some fraction of the population. Thus, while they may have been public in name, these utilities could not ordinarily rely on the general revenue generation powers of the state to cover exceptional or unanticipated expenses. Finally, in the European context to an extent unacceptable in the United States (with the notable exception of some municipal public utilities), a third group of 'public' landowners have been commonplace in many countries: state-owned enterprises. These have ranged from financial institutions to steel and coal producers and durable goods manufacturers. This latter group of public landowners have in many instances been in precisely those sectors most likely to have produced massive land pollution in the past due to ignorance.

Notwithstanding these legal distinctions, however, the countries we consider here all have to pay attention to the impact on current and prior *public sector special-purpose landowners* when allocating shares of financial responsibility and liability exposure across general government and individual landowners. In the same manner that placing all financial burdens on landowners for past pollution for which they may not have been responsible can undermine the economic viability of a business, such a pollution cost allocation could reduce the capacity of a public sector entity to deliver the services for which it was created. Such an impact inevitably produces a general cost to society: either required services are constrained and become less available and/or more expensive for users, or the special-purpose entity becomes dependent upon general government units for supplemental revenues, thus raising taxes for the economy as a whole. In suggesting dimensions that affect contaminated land policy, we can thus identify two distinct subdimensions.

Total public landownership, which is simply the proportion of land in the country held by the public sector or agencies it has authorized funded, at any level of government. This measures the extent to which the state as a whole may share liability for past pollution due to its landownership. The higher the proportion, the less significant the issue of the allocation of shares of financial responsibility between the national state and individual landowners, since the owners would become more and more coterminous with the state.

Sub-national public landownership, which provides a measure of the extent to which the public holdings are held by sub-units of the national state. Whether the landholder is a state, county or local government unit, an autonomous or quasi-autonomous non-governmental organization, or a service or goods-producing public or public–private entity, it benefits a subset of the national population. The issue of cost-sharing between the beneficiaries of the services of that arm of the state and the population as a whole becomes a greater policy concern as the sub-national ownership share grows.

ACCESS TO CAPITAL AND TO LIABILITY INSURANCE

Notwithstanding the internationalization of capital markets and the world economy as a whole, access to capital varies across domestic markets. Developers of different sizes operating in different local markets or countries will experience some variation in the availability of funds for projects, in terms of costs of capital, total funds available, or other conditions of lending. Capital availability in any country will vary with both lending practices in general and with the availability of insurance that provides protection against uncontrollable risks and uncertainties. Locally, access to capital may vary with domestic patterns of spatial reallocation of economic activity.

Private sector lending practices that prevail in national capital markets can have a profound effect on the rate of cleanup of brownfield sites. Developers rely to some degree on their ability to raise capital in order to pursue projects, so the conditions under which funds may be available to them affect their ability to assess the extent of past contamination, if any, and their capacity to conduct cleanups. Lending practices are shaped by both the legal obligations of potential financiers such as regulations governing bank or retirement fund operations and by institutional traditions and common practices that vary from country to country. Past experience in a country with risky or uncertain projects also affects willingness to finance similar projects in the future, as does the availability of public sector insurance for loans. We address these dimensions in order.

Reliance on collateral for loans may be shaped both by regulatory requirements and tradition. The practice of 'name lending' by merchant banks in Britain, for example, is effectively a zero collateral requirement, in which loan risk is assessed relative to the reputation of the borrower. Government requirements that financial institutions maintain minimum levels of collateral for commercial loans, such as exist in the US, preclude such reputational lending — and can constrain lending if discovery of contamination depresses the market value of the collateral held by lenders on past loans. Ignoring market segmentation as it affects demand for collateral, we define this dimension as the average percentage collateralization expected or required in national lending markets.

Centralization of financial institutions in a national market can affect the availability of funds regardless of collateral rules. The capacity to assess a given project in its local context permits a potential lender to consider more dimensions of the local real estate market and other idiosyncratic factors than would be possible with a standardized assessment protocol. A developer routinely takes such local data into consideration, so a financial institution capable of including the same variables in its lending decisions would be more likely to agree with the assessment of the financial gains to be made from any given proposed project. Thus the issue may be more one of the centralization of decision-making on lending than of the extent of formal legal centralization. It would appear that, to the extent that local branches or divisions of centralized banks or other financial institutions are provided flexibility in loan assessment and risk appraisal, capital for brownfield reclamation efforts would generally be more available.

Financial institution risk aversion, if sufficiently high and pervasive, could virtually preclude access to debt financing for contaminated land cleanups, since they are inherently risky. It may be argued that developers are inherently risk *takers*, staking their assets and borrowing capacities on expected future returns, while financial institutions are risk *avoiders*, as evidenced by a range of practices including demands for collateral, reliance on long reputational track records of borrowers, pursuit of loan guarantees, and the like. The extent of such avoidance, however, can vary from one capital market to another, and across different types of financial institutions. Under US law, for example, retirement fund administrators can be held personally liable for failures to adhere to a 'prudent man' approach to investment. As a consequence of these legal pressures to avoid risks the fund administrators may reject projects that may be acceptable to many insurance companies, banks or savings and loan associations (building societies). Whatever the legal and regulatory requirements governing the mitigation of the consequences of past pollution in any country, the higher the overall aversion to incurring risk on domestic projects by its private sector sources of capital, the more difficult it will be to reclaim and reuse contaminated land.

Recent experience with risky loans will affect the availability of capital from any set of financial institutions. The experience of institutional failures due to unanticipated risks or excessive risk exposure can thus constrain lending not merely by reducing the number of lending institutions but by inducing a temporary shift in the direction of higher risk aversion on the part of remaining potential lenders. This effect has been clearly evident in the United States as a consequence of the savings and loan débâcle of the late 1980s, which saw billions of dollars lost as institutions that had previously been constrained to investments in housing broadened their lending. Their failure to understand or cover the risks associated with the potentially higher returns to investment in non-housing projects led to the collapse of literally hundreds of such financial institutions. That failure has been blamed for capital shortages associated with higher levels of risk aversion on the part of commercial banks, insurance companies and other US financial institutions, scared by the widely publicized losses on loans. Obviously, then, the greater the level of such recent and well-publicized domestic loan losses, the harder it would be to finance contaminated land cleanups in any country. We cannot derive conclusions here about the extent to which the experiences of one group of potential financiers affect the behaviours of others, since the response is a matter of how the problems are perceived, a point to which we return in Chapter 7.

The availability of lending insurance in a capital market permits lenders to reduce their risks by sharing them with insurance coverers. Such insurance may be available from private sources or through public insurance pools. In some instances, the state pools may be jointly financed by contributions from the lending institutions themselves under legal requirements, as is the case with the US Federal Insurance Deposit Corporation and Federal Savings and Loan Insurance Corporation. In others, they may be provided at a fee to lenders but underwritten by the general revenues of the state, such as the loan insurance provided by the US Small Business Administration. In yet others private insurance availability may not be a significant issue. To the extent that a corporatist orientation dominates approaches to redevelopment, the state takes responsibility for cleanup costs and, in effect, indemnifies the developer and/or lender against excessive mitigation expenses, thus providing the effective equivalent of lending insurance.

The more readily such protection against risk is available, the greater the willingness of potential lenders to finance brownfield redevelopments. Obviously, to the extent that the insurance pools are underwritten by general state revenues, such insurance constitutes a process for generating what is ultimately public sector financing of projects that have potentially substantial private sector returns, so many of the allocation issues discussed above become germane to assessing the effectiveness and equity of cleanup policies.

General liability insurance practices and experiences affect the actual financial losses incurred by landholders or users under whatever national practices govern liability allocations for environmental damage or cleanup costs. The issue here is the coverage available to the parties which have caused the contamination rendered illegal under recent law. The US has had contentious litigation over the extent to which general commercial liability coverages in effect prior to the 1980 passage of CERCLA cover the new liabilities generated by the Act. Insurers have argued that their policies cannot retroactively be presumed to cover contaminated land liabilities for practices that were not illegal prior to the Act, while polluters claim that such liabilities are included in the unrestricted general policies that were in force. The broader the coverage of past pollution under such liability coverage, the easier it will be for the polluters to pay for cleanup of the lands they have contaminated since they are cost-sharing with their insurers.

Such practices, while of paramount importance in the US, a country reliant upon individualist private market processes, may not matter much in settings that rely more on state action. We are not aware of any analogues in European countries to the conflicts arising in the US courts over such liability coverages. We conclude that the corporatist acceptance of liability for the debts of past pollution in Europe reduce this factor to relative insignificance.

Experience of spatial displacement and capital flight may induce local political jurisdictions to attempt to minimize the private sector (and *local* public sector) costs of contamination. If localities — or even countries — fear capital flight from areas suffering extensive contamination, they may want to weaken regulations or otherwise facilitate redevelopment through improving the spatial competitiveness. The extent of such imperatives, which may be growing with the expansion of free trade areas and further internationalization of capital, may be related to a number of distinct factors.

Actual losses to external locales, whether of jobs, income, or tax revenues, will increase concerns for spatial competitiveness and thus tend to weaken regulations or reduce the cleanup standards imposed unless non-local finance for mitigation is available. Such losses may be across national boundaries (and may be accentuated by the development of free trade zones such as that of the European Economic Area or the North American Free Trade Agreement). The recent addition of Sweden, Finland and Austria to the ranks of EU members may be expected to result in pressure for higher uniform cleanup standards and stricter enforcement of European law, reducing the disparities that have been blamed for some migration of industry to countries deemed lax in enforcement, such as Greece and Italy.

Such losses, however, may equally be important across local political boundaries, such as between cities and suburbs in a broad conurbation or metropolitan area. To the extent that local governmental jurisdictions rely on taxes generated by businesses operating within their boundaries or on taxation of real estate (as is the case in the United States and Germany, for example), this local concern is accentuated.

The extent of land dereliction and abandonment, whether or not it was generated by avoidance of potential liabilities associated with contaminated sites, also heightens the importance of this factor. Older industrial areas that were the homes of industries that have declined in importance or moved to other locales, are especially liable to face such high dereliction levels. The US 'Frostbelt', encompassing most of the northeastern and midwestern states, Britain's Black Country and South Yorkshire, Germany's Ruhrgebiet coal and steel region and ex-GDR Bitterfeld area, France's Alsace-Lorraine, the Sambre-Meuse region of Belgium, and Spain's Asturias region are all examples of areas suffering levels of abandoned land and underutilized sites that are substantially greater than those in their countries as a whole. Such locations may be most seriously damaged by the new constraints on land reclamation and reuse generated by the concern over past contamination. Whether they deserve special treatment and/or subsidies from larger political jurisdictions for the costs they face is not important to us here. The existence of such exceptional cases places the whole issue of contaminated land policy in a different political context, that of interlocational competition for jobs, than would exist if there were a more even spatial distribution of contamination.

CURRENT POLICY AND POLITICAL PRESSURES

We have reviewed a variety of conditions, traditions, and practices which may shape or influence approaches to the reclamation, redevelopment and reuse of contaminated brownfield land. It may well be that positions on some of the dimensions we have considered always occur simultaneously — such as greater local flexibility in the presence of uneven spatial distributions of pollution or higher levels of state willingness to cover costs from general revenues with higher rates of public sector landownership. Thus this massive array of contextual variation factors may be reducible to a smaller set of key variables. Such a determination awaits the findings of Parts II and III, in which we examine the actual conditions and practices prevailing in the United States and Europe. We explicitly consider this reduction in Part IV of this book.

One facet of prior experience has, however, been ignored in our discussion thus far: the actual law and practice that has governed approaches to contaminated land cleanup since the issue first appeared on the political agenda in any country. It is here that the uniqueness of the United States becomes evident: the Superfund Law has been in place since 1980, so the US has a body of experience attesting to the strengths and weaknesses of its approach, which is unmatched by that of any other country. This experience is instructive because of the length of time the legislation has been in effect, because it is more comprehensive than the legislation present in any other nation-state, and because, given its size, the US illustrates the geographical and spatial issues that may face the European Union as it moves towards a comprehensive approach of its own. We do not mean to suggest that there are not lessons to be learned from the individual national experiences of the EU member-states — on the contrary, we use those examples to critique US policy and suggest improved approaches in our concluding section. However, the US policy is, itself, currently under review for potential (and probable) revision, hence the comparison of this single national case to that of the emerging transnational EU policy.

Experience under current and prior cleanup efforts shapes new proposals for change. Some experience may be negative, such as massive delays and litigation costs associated with liability allocation under CERCLA in the US. Other experience may be positive, such as the mitigations achieved under area-wide approaches in Northern Italy and in the Ruhrgebiet's IBA–Emscher Park efforts. Still other experiences may be deemed inappropriate or inequitable by some economic actors who then lobby for change to avoid liability or otherwise improve their economic prospects. Whether the issues raised by such interested parties are appropriate or not, and whether or not they warrant policy changes, the matters they place on the table further shape the political debate and may affect future regulatory approaches.

The strength of environmental movements and Green parties will obviously affect how either the past environmental problems or difficulties in operationalizing existing policies and making them effective and efficient will be interpreted in terms of political importance. No public policies emerge simply from existing *conditions*, however problematic they may be ecologically or in terms of human health. They enter into the political debate and appear on political agendas only to the extent that they constitute *problems* (Dye 1966; Lasswell 1971). The transmutation of conditions into political problems requires the presence of one or more political advocates (Dror 1971; Hofferbert 1974).

The political strength of popular or broad-based environmental movements, the existence of small groups willing to take disruptive action to publicize their concerns, and/or the electoral powers of environmental parties or candidates thus define the national level of attention paid to environmental issues in general, and contaminated land in particular. This strength may emerge from ideological roots, but is certainly enhanced by experience of adverse environmental consequences or policy failures. The form that the political influence takes may vary with the number of active political parties, the electoral and government process, and other factors that vary from country to country. However articulated, the relevant factor here is the political salience and visibility of 'environmentalism', broadly conceived. Without such an environmentalist concern — which is by now shared across the industrialized world — contaminated land would not be a political and public policy issue.

We are not considering here the presence or absence of an environmentalist ethos, but rather the influence and political power of the advocates of environmental causes. The scale on which we might measure this dimension, then, might be the proportion of the electorate in a country that deems the environment a matter worthy of national

Dimensions of National Contaminated Land Policy 51

or international policy intervention. The higher that proportion, the more likely it will be that land contamination and its mitigation will be closely regulated by the national state.

Now we turn to consideration of the experience on the ground, examining first the United States and then some member-states of the European Union as well as the Union itself. With evidence from detailed examinations of the actual experiences on both sides of the Atlantic, we can return to the dimensions introduced here to compare and contrast the approaches, successes and failures we have uncovered and draw some conclusions about context and process. Such findings are essential to our effort to suggest some at least partially generalized approaches to the contaminated land problem.

PART II

3. The Emergence of Environmental Concern and Toxics Policy in the United States

Public policy issues emerge from a political process, not simply from an existing condition or problem. Industry in the United States generated toxic wastes and contaminated land for at least a century before toxics became a political problem. We examine here the growth of the US environmental movement and the rise of a public concern with the potentially dangerous byproducts of traditional production processes and practices. A review of the growth of political imperatives to address land contamination is essential to an understanding of how the US approach to contaminated land has emerged and developed.

We begin with a brief discussion of the growth of US environmentalism since the 1960s. Next, we turn to the federal government response and provide an overview of the Environmental Protection Agency (EPA), the organization created in 1970 to take responsibility for environmental regulation. Third, we sketch the history of the main pieces of legislation which address hazardous wastes in general. Fourth, we turn to discussion of the grassroots struggles in response to discoveries of land contamination, which provided the political pressures resulting in the passage of the Comprehensive Environmental Response, Compensation and Liability Act (CERCLA) in 1980. Finally, we review the initial experience under CERCLA and the implementation experiences that produced amendments associated with reauthorization of the Act's taxing powers in the Superfund Amendment and Reauthorization Act of 1986 (SARA).

THE EMERGENCE OF US ENVIRONMENTALISM

The last three decades in the US have seen an increase in environmental concern and activism which has had a profound effect on legislation passed by elected officials. The emergence of the movement in the 1960s may be attributed to a number of distinct, but interacting, factors. These include information about ecological problems and human impacts on the global environment, specific instances of apparently acute damage to human health, and the affluence of post-war American society that permitted attention to be directed at quality of life in addition to simply quantities of income or material consumption levels (Dunlap and Mertig (eds) 1992).

Certainly, the 1962 publication of Rachel Carson's best seller, *Silent Spring*, which described the ecological consequences of using pesticides such as DDT in terms the popular press could easily grasp, provided a key catalyst. Moreover, the timing of the volume was serendipitous. The 1960s' counterculture had begun to develop from the struggles of the US Civil Rights movement in the early sixties, but not yet shifted its focus to the Vietnam war. Carson's book, then, developed an audience over the decade among politically active youth becoming increasingly critical of their society and inclined to take action to solve social problems. The values and beliefs of the counterculture were already centred on distrust of what President Eisenhower, in his farewell address, had labelled the 'military-industrial complex'. Criticism of private companies for their environmental excesses thus was willingly accepted.

While *Silent Spring* may have contributed to the passage of the Endangered Species Act (1966), the broader strength of concern for the environment, which was already extant, is evident in the earlier passage of a 1963 Clean Air Act and 1964 Wilderness Act. This legislation may be traced not only to increasingly visible urban smog, but also to growing public interest in outdoor recreational activity (Dunlap and Mertig (eds) 1992). Subsequent events in the decade also brought a broader range of environmental issues to the attention of the public and policy makers. In 1968, US astronauts brought photos back from the moon that sensitized the population to a small, finite and ultimately fragile earth. Two widely publicized incidents in 1969 capped the decade and solidified a base of popular support for intervention by the national state to protect the environment. January of that year saw the rupture of a Union Oil Company drilling rig platform in the Santa Barbara Channel, sending hundreds of thousands of gallons of crude oil onto the beaches. Later, that summer, a sailor threw a cigarette into Cleveland's Cuyahoga River and the river burst into flames.

Both events provided ample pictorial evidence for both print and electronic media and resulted in further broadening of the political base for environmental regulatory policy. On 1 January 1970, President

Emergence of Environmental Concern and Toxics Policy in the US 57

Richard Nixon signed the National Environmental Policy Act, placing broad obligations on the national state and its agencies to incorporate questions of environmental impacts in its decision-making processes. By 22 April 1970, when the country observed the first declared 'Earth Day', participants included members of the US Senate and House of Representatives as well as thousands of students on college campuses, and millions of others in communities across the continent. President Nixon issued an Executive Order creating a special independent administrative agency to address environmental problems, and, with the concurrence of Congress, the Environmental Protection Agency (EPA) was formally created on 9 September. Environmentalism thus became solidly institutionalized in the United States.

The following decade saw over two dozen new pieces of environmental legislation passed by Congress, some in direct response to international events such as the United Nations Conference on the Human Environment in 1972 and two OPEC oil embargoes, which raised concerns about non-renewable energy resource consumption. These included a new Clean Air Act (1970); Water Pollution Control Act (1972); Environmental Pesticide Control Act (1972); Ocean Dumping Act (1972); Endangered Species Act (1973); Safe Drinking Water Act (1974); Resource Conservation and Recovery Act (1976); Federal Land Policy and Management Act (1976); Toxic Substances Control Act (1976); Clean Air Act (1977); Surface Mining Control and Reclamation Act (1977); Federal Insecticide, Fungicide, and Rodenticide Act (1978); and the National Energy Act (1978). This period of unprecedented environmental concern culminated in President Jimmy Carter's signing of CERCLA in December 1980.

Support for environmental issues has continued to rise. As of 1990, over half of the US public (54 per cent) felt that environmental protection laws had not gone far enough. The percentage of people who believed that their local environmental quality was getting worse rose from 34 per cent in 1983 to 55 per cent in 1990. The percentage who believed that most or many underground water sources were contaminated rose from 28 per cent in 1981 to 54 per cent in 1988. Results indicate continuing support for environmentalism; in 1990, a remarkable 73 per cent of the public considered themselves to be environmentalists (Dunlap 1992).

By the 1990 twentieth anniversary Earth Day celebration, membership in the 12 major national environmental organizations totalled over three million, a substantial increase from their 123,000 members in 1960 (Mitchell et al. 1992). While many of the organizations date to the early decades of the twentieth century and their membership remains under two per cent of the US population, they have become potent political forces through their participation in electoral campaigns, congressional and administrative lobbying, and litigation. The key political actors with respect to these activities include the Sierra Club, National Audubon Society, National Parks and Conservation Association, Izaak Walton League, Wilderness Society, National Wildlife Federation, Defenders of Wildlife, Environmental Defense Fund, Friends of the Earth, Natural Resources Defense Council, Environmental Action and the Environmental Policy Institute.

The impact of these organizations on public attitudes is perhaps best measured by the extent to which environmental protection has risen in importance to challenge the primacy of economic growth. When asked 'Which of these two statements is closer to your opinion: We must be prepared to sacrifice environmental quality for economic growth. We must sacrifice economic growth in order to preserve and protect the environment', 38 per cent of respondents in 1976 were prepared to sacrifice economic growth and 21 per cent the environment (Cambridge Reports, Inc. 1986). By 1990, 64 per cent supported sacrificing economic growth for environmental protection and only 15 per cent considered economic growth to be more important (Cambridge Reports/Research International 1990). Given the recessionary conditions in the US economy in 1990, this is a spectacular expression of concern.

THE ENVIRONMENTAL PROTECTION AGENCY

All levels of government in the US, national, state and local, are now involved in some capacity in environmental management. Most states have several agencies involved in environmental protection (e.g., a Department of Health, Department of Water Resources, Department of Natural or Environmental Resources). At the local level, there are a variety of environmental quality management districts and agencies in cities and counties, many mandated by, or involved in regulatory practices required to conform to, state and federal requirements.

It is noteworthy, however, that the majority of these sub-national public sector efforts have been, until recently, focused almost exclusively on air and water pollution problems. Such issues are amenable to state and sub-state regulation in part because they can be addressed within local watershed districts and air basins. Other environmental problems involve either more complex movement of pollutants or are not narrowly geographically confined (Ringquist 1993).

At the federal level, many organizations deal with environmental issues. For example, the Department of the Interior, founded in 1849, has extensive responsibilities over environmental and land management. It houses the National Park Service, the US Fish and Wildlife Service, the Bureau of Land Management, the Minerals Management Service, the Office of Surface Mining, the US Geological Survey, the Bureau of Reclamation and the Bureau of Mines, all of which long predate the wave of post-1962 environmental laws.

The Environmental Protection Agency (EPA) is thus a relative newcomer to the federal environmental bureaucracy. It is, however, the only regulatory agency that reports directly to the President. All other regulatory powers are wielded by units of federal departments that report to Cabinet-level Secretaries. Its 1970 creation was a political response that involved consolidation of narrow pollution regulatory activities from a variety of different departments into a single, more independent, agency. The Agency is thus, in theory at least, capable of better integrating its promulgation of regulations for implementation of environmental laws to minimize conflicts and inconsistencies and to facilitate compliance and regulatory enforcement.

Organizationally, the Agency includes divisions of water, solid waste, air and radiation, pesticides and toxic substances, all of which have specific regulatory obligations and powers and offices of enforcement or compliance monitoring. Both to permit administrative decentralization and to facilitate specialization in the problems, policies and practices of the individual states, the Agency operates ten Regional Offices, roughly corresponding to US Census regions. Overall responsibility for enforcement of environmental regulations is vested in an Assistant Administrator for Enforcement and Compliance, who is responsible for both civil and criminal litigation and prosecutions. The Agency's Office of Criminal Enforcement employs personnel with experience investigating criminal activity who act as environmental law-enforcement officers, collecting evidence and serving warrants with full police powers (Smith 1992).

The EPA also serves as a lead agency for the conduct of research on environmental problems and their mitigation. This function not only integrates research activity that might otherwise be fragmented across a number of different federal departments, but serves the needs of the states and other smaller jurisdictions for scientific evidence to guide promulgation of policies and standards. Sub-national environmental policy is also enhanced by deadlines and sanctions for local-level non-compliance with area water and air quality standards which stimulate regulations and regulatory enforcement by lower levels of government (Ringquist 1993).

Overall, then, the EPA serves as a central coordinative body for environmental protection. It bridges national government administrative divisions while it integrates the policies and practices of the various levels of government in the US federal system. In its absence, the mass of US environmental legislation would, arguably, be far more burdensome for those entities that would be forced to conform to regulatory requirements administered by an array of administrative arms of different departments and bureaux.

THE REGULATION OF HAZARDOUS SUBSTANCES AND WASTES

Toxic and hazardous substance use, transport and disposal are regulated by more than a dozen different federal agencies under a variety of federal acts including the Toxic Substances Control Act of 1976 (TSCA), which regulates the introduction and use of new hazardous chemicals; the Federal Insecticide, Fungicide, and Rodenticide Act; the Food, Drug, and Cosmetics Act, the Clean Air Acts, the Clean Water Act, the Occupational Safety and Health Act, and the Safe Drinking Water Act (Smith 1992).

However, two federal statutes are most important in regulating the treatment and disposal of hazardous substances. These include CERCLA and the Resource Conservation and Recovery Act (RCRA) passed four years prior to CERCLA in 1976. RCRA addresses current activities involving the disposal of toxic substances while CERCLA was subsequently passed to correct the damage done to the land by prior unregulated or illegal disposal of hazardous chemicals and contaminated goods and equipment.

Prior to RCRA, which accompanied TSCA on the congressional agenda, only air and water and not land were protected from hazardous disposal. The Act requires a permit in order to store, treat, or dispose of hazardous materials. Only firms that demonstrate that they have the financial resources, insurance and the expertise to manage toxics are granted these permits. RCRA gives the EPA the power to hold individuals criminally liable and to impose fines on companies for improper disposals. At the time, it was deemed to have closed 'the last remaining loophole' in environmental laws designed to protect the public (Barnett 1994:57). However, RCRA has been plagued with enforcement problems: as of 1986, close to 80 per cent of the disposal sites actually licensed by RCRA were in violation of its requirements (Bloom 1986).

RCRA was signed into law by President Gerald Ford in October 1976. The EPA efforts to write the implementing regulations were fraught with contentious debate, primarily over designating which substances (and what amounts of them) were hazardous. The EPA's 1978 draft regulations, which had to be announced under the law's implementation requirements, were met with considerable opposition from both industry and environmentalists. With no definitive proof or evidence of toxicity and/or appropriate standards for levels of human exposure, industry advocates claimed the regulations covered too many substances and were too restrictive; environmentalist organizations argued the opposite, that too many substances were not covered or exposures permitted were too great. As a result, the first substantial set of final regulations for implementation of the 1976 Act were not published until May of 1980, just seven months before the passage of CERCLA (Landy et al. 1994).

During the regulation development period, public concern grew over the health effects of contaminants found in factories and communities throughout the country. This concern may well have been an unintended byproduct of the intense industry lobbying efforts challenging the scope of RCRA requirements. As environmentalists built their cases by pointing to instances of contamination or workplace exposures with apparently serious consequences, they, directly and indirectly, may have stimulated local grass roots concerns. These worries, augmented by a small number of widely publicized cases of *past contamination*, a loophole or remaining problem *not* closed by RCRA, led to the passage of CERCLA.

The 1970s' Development of Concerns over Toxics

Toxics pervade the US economy. It has been estimated that Americans are responsible for generating more than one metric ton of hazardous waste per person per annum (US CBO 1985). Obviously, toxic chemicals were widespread long before the passage of RCRA, the four-year debate over the promulgation of implementing regulations, and the passage of CERCLA. However, the issue did not attract public attention easily since chemical contamination of land and ground water, unlike air and surface water pollution, is not visible.

There is virtually no evidence of widespread public or environmental movement concern over hazardous wastes and other substances prior to the mid-1970s. By 1990, however, 65 per cent of general public respondents to a Roper survey rated the danger of abandoned hazardous waste sites as 'very serious', placing them ahead of nuclear accidents, contaminated drinking water and other environmental problems listed (Roper Organization 1990). The emergence of this concern is largely traceable to media coverage of some key exposure cases.

Love Canal, which first gained widespread media attention in August 1978, was the most notable of these situations (Levine and Levine 1982; Brown 1981). A problem was first uncovered in 1976, when pollution discovered in Lake Ontario was traced to Love Canal. Further investigation of what first was considered a water pollution issue led to the involvement of local, state and federal officials, including the EPA. The basic facts were simple: the abandoned canal, located in Niagara Falls, New York, was used by the Hooker Chemical Company between 1942 and 1952 to dispose of more than 21,000 tons of chemical wastes. When it was full, the canal was covered with earth by the company and then sold for one dollar (effectively donated) to the local school board. The property was first used for a school; eventually a residential neighbourhood developed on and around the canal site.

Studies of reported birth defects and child-health problems among residents at Love Canal led New York State Health Commissioner Robert Whelan to declare a 'public health emergency' in August 1978. He ordered further studies and provided guidance to residents on actions they could take to protect themselves, advising pregnant women and young children living on streets adjacent to the canal to temporarily leave the area. Whelan's subsequent report in September, *Love Canal: Public Health Time Bomb*, depicted the site as a 'devastating, modern day disaster' and a 'great and imminent peril' (1978:1,30).

Both the declaration of an emergency and the phrase identifying an imminent peril were terms legally required to release funds for studies and response actions. The language, however, was extremely volatile and elicited broad response and publicity. Electoral politics also broadened the attention given the issue, as Governor Hugh Carey of New York State was up for reelection in November 1978. He created a state task force and agreed that 239 families living on two streets adjacent to Love Canal would be relocated at state expense. These actions placed a gubernatorial seal of approval on the validity of the environmental damage claims. The Federal Disaster Assistance Administration was called. President Jimmy Carter shortly thereafter declared an emergency, helping out Governor Carey, his fellow Democrat.

Governor Carey became more cautious later after he was reelected, when he was informed that the federal government would not pay for polluted properties and further studies indicated that the original research findings on health risks were tenuous. Residents were informed in 1979 that not all families who wanted to be relocated would be. This decision was interpreted in the media as government insensitivity. At the same time, the EPA launched a series of lawsuits against the Hooker Chemical Company. A study indicating adverse health effects was introduced and findings leaked to the press before the research could be evaluated. During May and June, the story was in newspapers and featured on TV news programmes across the country. The EPA itself later disavowed the evidence originally introduced, but by then public opinion had rigidified. A panel of experts commissioned by Governor Carey found previous studies at Love Canal to be seriously flawed, but the panel's conclusions had little effect on public opinion. On 19 May, Love Canal residents held two EPA officials hostage. Eventually, over $15 million was spent to relocate a total of almost 1,000 households (Barnett 1994).

Other cases also made national headlines. In 1979, concentrations of carcinogenic compounds found in private water wells in Woburn, Massachusetts, were linked by a citizens' action group to high instances of leukaemia cases in the area (Brown and Mikkelsen 1990). Hardeman County, Tennessee, became known as 'The Valley of the Drums' when drinking water was found to be contaminated by the leaching of 300,000 drums of toxic waste that had been crushed in unlined trenches over the course of a decade of unregulated dumping (Finsterbusch 1989). News of these cases, which followed in the wake of the Love Canal publicity, received attention in newspapers and television news programmes and testimony from community members was drawn upon at hearings on the passage of CERCLA.

These public reactions could have been anticipated. Social psychological research in the seventies had already documented that the 'pernicious character' of chemical contamination tends to generate distinctive and exceptional dread in the US (Combs and Slovic 1979;

Lichtenstein et al. 1978; Slovic 1979, 1992; Slovic et al. 1976). A risk is perceived to be more threatening and less acceptable when it is unobservable, unknown, and delayed in its manifestation of harm. Toxic chemicals provide the quintessential example (Slovic 1992). Chemical pollutants *per se* have an especially menacing aura in US culture that evokes exceptional anxiety. Chemicals found on brownfield sites may evoke particular apprehension because their origins, content and effects are often unknown. Indeed, there are over 70,000 chemicals currently in use in the US, but the EPA has assessed the risks of only two per cent of them (US GAO 1994e). Issues of fairness also shape conclusions regarding risk acceptability: risks are perceived to be less tolerable when they are judged to be unjustly imposed rather than voluntarily accepted (Starr 1969).

Case studies of communities found to be contaminated support these findings (Brown and Mikkelsen 1990; Edelstein 1988; Hallman and Wandersman 1992; Picou and Rosebrook 1993; Vyner 1988). Those who believe they have been exposed to toxins known to be present experience anger, frustration, helplessness, depression and intense anxiety, none of which are appeased by assurances from toxicologists who, themselves, may be uncertain or disagree about their conclusions regarding the effects of the chemicals. Residents who voice concerns over the pollution may be ostracized by neighbours who stigmatize them as carriers of a mysterious disease or who fear that news of the contamination will jeopardize jobs and reduce property values.

In the light of what we know about risk perceptions, then, it is not surprising that since the late 1970s there has been a dramatic increase in local struggles, which increased anxieties about chemical contamination. Local grassroots organizations formed in response to findings on or fears of environmental health hazards that directly affected them, such as toxic waste disposal sites and contaminated water. Unlike mainstream environmentalists, members of these groups are more likely to be working class and minorities, primarily because the health hazards are disproportionately located in their communities (Bullard 1990). Such local health self-defence efforts have broad media appeal, so politicians cannot easily ignore communities enraged by contamination, a growing proportion of the US electorate.

The Coming of Superfund

The US Congress that authorized Superfund in 1980 was elected the same year that Governor Carey won reelection in New York as Love

Canal gained national prominence. However, CERCLA itself did not originate with the Love Canal experience, for all that the publicity may have shaped the final Act. EPA officials had already begun drafting the legislation in 1978, and submitted a bill to Congress two months before the major news stories broke on the Love Canal site.

Landy et al. (1994) contend that Douglas Costle, the EPA administrator at the time, was building a public health rationale for EPA in order to stabilize and increase its funding and scope of authority. Under his administration, protection of public health replaced ecological preservation as EPA's primary concern. Costle took advantage of the furore caused by Love Canal; he even conducted a systematic search to uncover new contaminated land cases in each EPA region. Thus, at the time of the congressional hearings on CERCLA, EPA was able to warn representatives of other 'ticking time bombs' located in their own back yards.

Notwithstanding the publicity over cases of land contamination, EPA's original draft of the Superfund legislation took seventeen months to get through Congress and passed just before Congress adjourned in December 1980. For the most part, the extensive hearings produced publicity for legislators and the EPA, but did not substantially alter the basic aspects of the draft legislation. Two aspects of enforcement capacity were removed, however, at industry insistence. First, a victims' compensation provision which gave those damaged by contamination the right to sue polluters in federal court was eliminated, weakening private action potential (perhaps because the chemical industry feared lawsuits brought by community-based organizations). Second, at the insistence of the oil and gas industry and their lobbyists, EPA was not granted authority over spills that affected groundwater, such as oil spills (Barnett 1994).

The underlying rationale of CERCLA was the polluter pays principle, that is, creators of pollution should pay for the damage they do. Although RCRA authorized the government to order cleanup of hazardous waste, the problem was that many sites had long since been abandoned. Many past polluters no longer existed as corporate entities or did not have the financial capacity to pay. To address this problem, CERCLA established a revolving $1.6 billion five-year fund financed mainly by taxing petroleum products and chemical feedstock (the true 'Superfund', which we designate here as the 'Fund'). The moneys were to permit immediate efforts to remediate properties listed on a national priority list (NPL) of severely contaminated or acutely dangerous sites, anticipating recovery of costs from responsible parties at a later time, that is, 'shovels first, lawyers later'.

CERCLA's 'lawyers later' provisions incorporated specific liability standards to ensure that those held responsible for a release would pay the full cost of cleanup. If they refused, the EPA could sue them in federal court and obtain treble damages (a standard derived from US antitrust legislation). Determination of the appropriate shares of cost to be borne by different parties was made a problem for the defendants, not the EPA or the government. In order to assure maximum availability of non-federal funds for cleanup, parties found to be responsible would be exposed to strict, joint and several liability, provisions we address in depth in Chapter 4.

The Initial Implementation of CERCLA

The Superfund bill was signed into law on 31 December 1980, as virtually the last domestic act of the Carter Administration. President Ronald Reagan, an avowed opponent of government regulation of the private market economy, took office in late January 1981. The new President employed three tactics to limit federal environmental regulatory activity: (a) the appointment of anti-regulationists to key posts; (b) executive sector reorganization to allow his loyalists to control the EPA, regardless of the intents of its civil service cadre; and (c) budget constraints on the agency (Landy et al. 1994).

Anne Gorsuch-Buford, the new EPA administrator, was a vehement Colorado opponent of hazardous waste disposal regulation, Superfund programme head Rita Lavelle came to Washington, DC from a corporation that EPA had found to be the third worst polluter in California, and other appointments reflected similar opposition to government controls (Barnett 1994). The EPA Office of Enforcement was abolished and its staff distributed among various programme offices, with the expected results: between 1980 and 1982, the number of cases the EPA referred to the Justice Department declined by 50 per cent and the number of enforcement orders issued dropped by one-third, despite the passage of a major new regulatory act, CERCLA. The budget was also cut from $701 million to $515 million between 1981 and 1983, and the number of authorized staff positions declined 22.6 per cent. In fact, by 1982, a space of little more than a year, close to 40 per cent of the Agency's workforce had left due to layoffs and to resignations (Landy et al. 1994; Barnett 1994). The legal evisceration of the Agency was a success, and Superfund was, as might be expected, a failure in its early years, as measured by progress in identifying and cleaning up contaminated sites.

The implementation of CERCLA was further impeded by illegal acts by the EPA Administrator and her staff. Conspiracy with polluters, channelling of cleanup funds to locations to assist Republican candidates for congressional elections in 1982, and other abuses were uncovered by the (Democrat-controlled) House Committee on Energy and Commerce, forcing the (Republican) Justice Department to investigate the Agency (Barnett 1994; Landy et al. 1994). Congressmen concerned about slow cleanups at sites in their districts subpoenaed EPA records and charged both Administrator Gorsuch-Buford and Assistant Administrator Lavelle for the lack of progress in site remediation, accusing them of both misconduct and incompetence.

The US House of Representatives voted Gorsuch-Buford in contempt of Congress for failing to furnish documents; she resigned in March 1983. Lavelle, accused of obstructing a congressional investigation, was convicted of perjury in December 1983. Three years of potential progress on the contaminated land problem were lost, however, along with some of the political and judicial momentum generated in the late seventies, as the Superfund was starved by the request for and provision of only one-eighth of the moneys authorized by Congress (Barnett 1994).

The Coming of SARA

The delay was politically important, as the original CERCLA law provided for only five years' appropriations to the Superfund. SARA, the Superfund Amendment and Reauthorization Act, was moving onto the political agenda even as the EPA's original Nixon-appointed Administrator, William Ruckelshous, returned to pick up the pieces left by President Reagan's original appointees. Four factors set the stage for the SARA reauthorization debates: (1) Reagan's anti-environmentalism; (2) the EPA scandal; (3) publicity about new contaminated land struggles; and (4) an increase in broad-based environmentalism, stemming, in part, from the first three factors. The first considerations were discussed above, so we turn now to elaborating on the latter two.

The early 1980s saw an increase in the number of community-based environmental protection groups, particularly in minority areas with inordinately high levels of toxics exposure and a history of civil rights struggles in their communities (Bullard and Wright 1992). Residential areas around the country were looking at their commercial and industrial neighbours with new, fearful eyes. Their fears were, in fact, stimulated by the very existence of the Superfund programme, despite delays in its implementation. One chemical alone, dioxin, led to the federal EPA designation of thirty-one sites as contaminated in the Summer of 1982 (Barnett 1994).

Primary among these cases of contamination was Times Beach, Missouri, a small town of 2,200. Practices known to produce widespread pollution with dioxin had been studied in Missouri since 1971. The problem had risen and fallen in significance as the scientific findings on the danger of dioxin contamination changed over the decade (Barnett 1994). In 1983, all Times Beach residents were evacuated from their town by the EPA (at a cost of some $35 million). A decade earlier, waste oil laced with dioxin had been sprayed on the unpaved streets to keep the dust down (Piller 1991). That chemical contamination was, by 1982, deemed to present an extreme danger. The town disappeared in months and is now fully abandoned.

The urgency of cleanups as well as the failures of the EPA were on the political agenda as the taxing authority for Superfund was due to expire at the end of 1985. The House and Senate each passed their own Superfund reauthorization bills but they were unable to reconcile their differences, especially in the face of Reagan Administration opposition to the Act. Thus, beginning in 1985, Superfund authority lapsed for just over a year and the programme's efforts were once again significantly scaled back; projects were put on hold until Reagan signed a revised compromise bill into law in October of 1986.

Part of the political problem facing Superfund advocates was the change in the character of the environmental movement and its objectives that took place in the late seventies and early eighties. While the community-based organizations grew apace, mainstream environmentalism declined somewhat in the late seventies, as people were lulled into thinking that environmental problems were being solved under the Carter Administration (Dunlap 1992). The anti-environmental orientation of the Reagan Administration reversed this trend and revitalized environmental organizations. However, this regenerated environmental movement was distracted in part by growing concerns for more global issues than local land contamination.

National environmental organizations were able to mobilize support in the earliest days of the Reagan Administration. From 1980 to 1983, memberships in national environmental lobbying organizations surged: the Sierra Club grew by 90 per cent and the Wilderness Society grew by 144 per cent (Mitchell et al. 1992). One national survey found that those who felt that environmental laws had not gone far enough rose from 33 per cent of respondents in 1980 to 48 per cent in 1983 (Dunlap 1992). Environmental concern was revitalized by warnings over ominous global ecosystem problems including global warming, ozone depletion, acid rain and loss of the world's rain forests. These issues shifted the focus from the local to the international, however, a pattern supported by much of the institutionalization of environmental awareness in schools.

Yet another factor complicated the SARA debates: the Democrats' concern for ensuring that an anti-environmentalist or anti-regulationist Republican Administration could not simply refuse to enforce the laws, as the Reagan EPA had been doing in practice. For example, under Gorsuch and Lavelle in early 1983, the Reagan Administration either suspended RCRA regulations or otherwise rendered them ineffective, for instance, regulations covering incinerators.

The previously promulgated ban on dumping bulk liquids into landfills was suspended in February 1982. Public and congressional criticism was so intense that the ban was reimposed less than a month later. In 1984, Congress amended RCRA to create new means for controlling what it perceived to be an irresponsible agency. The amendments deployed a device, called a 'hammer', which temporarily deprives an agency of its rule-making authority when it fails to meet a specified deadline and imposes a statutorily defined rule instead.

By the time Congress began the reauthorization process in 1985, it realized that the problem of contaminated sites was substantially worse than it was estimated to be in 1980. The total cost of the Superfund programme was estimated at the time to be $100 billion.

Recalcitrant management of the Superfund programme under the Reagan Administration played a key role in determining SARA provisions. Previous congressional legislation defined objectives, leaving development of specific regulations to the EPA. However, public demands for environmental protection (especially in the aftermath of Bhopal, India, in 1984) together with Reagan's preference for inaction on environmental issues and his attempts to weaken the EPA, led Congress to take a more direct role in regulation (Barnett 1994). SARA therefore dictated schedules for the commencement of cleanups and limited administrative discretion, for example, by stipulating that permanent, rather than temporary, remedies *had* to be employed at Superfund sites whenever possible. It is ironic, however, that in the face of Reagan Administration resistance, the more environmentalist Congress still had to temporarily undermine the

Superfund by a lapse in its taxing authority in its efforts to protect it from administrative branch tampering.

Other provisions were added under SARA to induce more rapid settlements including mixed funding, a practice by which the government can assume part of the cleanup costs; non-binding assessments of responsibility, a procedure that allows the EPA to prepare non-binding allocations of cleanup costs; and *de minimis* buyouts which allowed the government to accept a fixed sum from small contributors of waste in exchange for an end to their liability. Funding for the Superfund programme was increased to $8.5 billion. Community activists won a victory with a right-to-know stipulation requiring industries to report on their releases of hazardous materials to a publicly-accessible nationwide information system that the public can access (Rich et al. 1993). The resulting 'Toxics Release Inventory' quickly produced major reductions in releases, as producers sought to avoid adverse publicity. Again, however, provisions for compensation to victims affected by contamination were excluded.

A footnote to the history of land contamination regulation remains, a conundrum that haunts the pursuit of certainty and cost-minimization in cleanup efforts. On the basis of data analysis of dioxin a decade after it was declared to be one of the most toxic chemicals known, an official of the federal Center for Disease Control who had been central in the 1983 decision to evacuate Times Beach, MO, claimed that the dioxin present had not posed a serious threat to human health (*New York Times* 1991). However, the debate over dioxin remains acerbic (Barnett 1994). In fact, a forthcoming EPA document reportedly provides indication that the chemical is more dangerous than was previously thought (Environmental Defense Fund 1994). The very uncertainty of the scientific findings fans the flames of the anti-contamination movement and strengthens citizens' community organizations for promoting cleanups of known disposal sites and other contamination. It simultaneously explains investors' virtual panic reaction to the possibility of their involvement in any brownfield site that might have been contaminated, however minimally, at some time in the past.

4. The CERCLA Experience and Debates over Change

> As we go about making the necessary reforms to the Superfund programme, we must keep our eyes focused always on what is really at stake for the health of our citizens....Cancer clusters, severe or even fatal birth defects, miscarriages, respiratory and neurological problems, and all the financial and emotional trauma of being continually exposed to the chemicals that are causing these problems. (Senator Frank Lautenberg of New Jersey 1993:3)

CERCLA taxing authority for financing its Fund through a special levy is subject to the US 'sunset' provisions, requiring periodic reauthorization. The impending lapse of taxing capacity at the end of 1995 and the desire to use the required legislative action to improve the Act stimulated extensive review of the experience with the Superfund. Evaluative efforts and proposals for revisions and amendments to the Act in the 1993–94 period provide a rich lode of data for assessment of the experience under CERCLA.

Decisions about cleanup standards and practices, as well as about legal liability and shares in the cleanup costs, will have a substantial impact on private sector and local government investment capacities. To the extent that changes in Superfund implementation can reduce unnecessary costs or speed the rate of cleanups, all the liable parties would benefit. As the most visible and widely publicized land pollution reclamation effort, CERCLA tends to influence other regulatory and cleanup practices, so policy modifications will influence the magnitude of the total cost figure for all hazardous waste treatments and removals, not merely those associated with national priority list (NPL) remediations.

We describe here key elements of the ongoing debate over the successes and failures of Superfund, beginning with a review of past NPL cleanup efforts, financing of the Fund, and the impacts of current liability principles employed in allocating cleanup costs. Next, we turn to discuss non-NPL sites and the effects of CERCLA on urban economic development. Those impacts helped generate different responses at the level of the individual states, summarized in the section that follows. We then briefly describe the changes in Superfund proposed in 1994 by the Clinton Administration in the 'Superfund Reauthorization Act', on which Congress held hearings but did not vote. The chapter ends with examples of successful cleanups, illustrating the Superfund process at work at non-NPL sites.

NATIONAL PRIORITY LIST CLEANUPS

CERCLA may be credited with considerable accomplishments in terms of cleansing NPL sites or those properties judged to constitute the most serious environmental threats. These achievements have been made despite the difficulty of the task and the initial years of the programme when progress was undermined during the Reagan Administration, including the funding hiatus of 1985–86. As of 1993, 126 sites in which cleanup measures were deemed necessary had been listed as construction-complete or deleted from the NPL (US GAO 1993). Thus, substantial amounts of hazardous waste threatening human health and the ecosystem have been removed or, at least, temporarily controlled. Despite these gains, NPL remediations have been criticized on several grounds including costs, prioritization of sites, cleanliness standards and lack of community involvement.

Costs of NPL Remediation Projects

After fourteen years of discovery of contaminated properties stimulated largely by the Superfund programme, the magnitude of the problem appears immense. While comprehensive inventories of hazardous waste properties have not been completed, a major, in-depth study conducted by the University of Tennessee estimates numbers and expenditures for non-federal NPL sites to range from 2,100 sites costing $106 billion to 6,000 sites totalling $302 billion (Russell et al. 1991).

Massive as these expenditures appear to be, they pale in significance relative to spending on air and water pollution. Projected annual costs for air pollution mitigation are $44 billion, and water programmes amount to $64 billion. In addition, about one-third of US land is federally owned and is managed by agencies including the Departments of Defense, Energy, Agriculture, and Interior. Defense and Energy Department cleanups pose spectacular problems, including explosives, chemical and biological warfare agents and, of course, nuclear waste. Federal Department of Energy facility cleanups alone could cost $360 billion (US GAO 1994a). The University of Tennessee study estimated that, under current policy, hazardous waste remedial costs, including Superfund, RCRA corrective action, underground storage tanks, Departments of Energy and Defense and state/private programmes will total between $478 billion and $1.05 trillion. Thus, Superfund itself is but a small part of the total environmental cleanup burden under which the US economy is labouring. Writing in 1991, environmental economist Paul Portney estimated, for example, that Superfund would comprise only 22 per cent of the cumulative bill for hazardous waste remediation (Portney 1992).

Prioritizing Cleanups

Comparative risk assessments are employed to prioritize NPL sites for cleanup. However, the process and logic of the procedures are intensely controversial among scientists and the public at large; environmental pollutants and their dangers may be either exaggerated or underestimated (US GAO 1994e). Assessing the risks of polluted properties is especially complex because a variety of toxins are often present and because of the unique geological features of each site — there is no 'average' Superfund site.

A deeper problem exists in the utilization of any risk assessments under the Superfund programme: the formula for combining the assessments of risks associated with different environmental media. The Hazards Ranking System (HRS) was especially developed for EPA to use in identifying potential Superfund sites. Originally, it was structured to combine risks associated individually with air, groundwater and surface water on a 0–100 scale. Despite the fact that CERCLA is focused on site contamination, *land* pollution and risks associated with it was not added to the formula until 1990. The formula is so structured that, in the presence of one high-risk polluted medium, another adds little to the overall total. Thus two high-risk sites may have similar rankings, even though one poses risk only through groundwater while the other poses high risks through two or more media.

The flaws in the arithmetic of the HRS have been compounded by the manner in which it has been employed. First, an HRS score of 28.5 out of a possible 100 was selected early in the programme on a purely arbitrary basis, in order to have the legislatively required 400 sites on the NPL (Hird 1994). This cutoff reflects relative risk differences, but has been attacked since it does not deal with *actual* risk. That is, it is possible that serious human health or ecological damage risks exist at lower scores, or that no such dangers exist at that particular level and the cutoff should be higher. Second, having been shorthanded, the EPA regions have, to a large extent, left full risk assessments incomplete, stopping work on sites as soon as it was clear that they posed risks above the 28.5 cutoff level. Thus, the HRS, flawed as it is, has not been available for prioritizing among the sites on the NPL (US GAO 1994e).

Even if the data were readily available to rank sites on the risks they pose, it is still not clear that the current 'worst-sites-first' approach (which replaced the 'spread-the-fund-thinly' logic that prevailed in the early 1980s) is an appropriate prioritization logic. It has been argued that, given budget constraints, the objective of reducing the most risk with the resources available might best be accomplished by cleaning a greater number of less risky sites, since the worst sites are also the most expensive to mitigate (Hird 1994).

Cleanup Standards

Determining the standards for cleanups has also been a matter of recurring concern. The law does not specify a standard level of cleanup nationwide. Neither the legislation nor its implementing regulations specifically address the question of 'how clean is clean?' (Edison Electric Institute 1988). The EPA has been criticized for not establishing cleanup goals at NPL sites and thus not achieving uniform standards of cleanliness. A GAO investigation of treatment plans made between 1987 to 1990 found that 11 of 18 plans studied did not include goals for all major contaminants (US GAO 1992).

The mitigation standards required under CERCLA and SARA must conform to 'applicable or relevant and appropriate requirements' (ARARs). These requirements are imposed by other federal legislation and, most significantly, by state environmental laws that may impose far more stringent cleanup standards. The use of ARARs may significantly slow cleanups because regulators disagree over their interpretation and because state legislation continues to get stricter. In the absence of ARARs, which generally have not been promulgated for soil, cleanup levels are determined on a site-specific basis.

In principle, satisfactory remedial alternatives are then evaluated using criteria including permanence, implementability and cost. In practice, remedy selection often appears to depend on additional factors including community acceptance and the availability of PRPs to share the costs (Baes and Marland 1989). While SARA mandated that the EPA seek permanent treatments whenever possible, such remedies are considerably more expensive. For example, the cost of *in situ* incineration of polluted soil is five times greater than capping the soil (Gupta et al. 1994). The question of whether the current cost reduction is worth the reduced health benefits and the restrictions on future economic development of a site due to residual contamination remains unresolved.

Need for Greater Community Involvement

SARA requires EPA solicitation of community input into remedial action plans at NPL sites and provides for Technical Assistant Grants of up to $50,000 to communities to allow citizens greater participation in cleanup decisions. Many residents, however, have criticized the programme for offering inadequate opportunities to participate fully in the decision-making process, particularly with respect to determining remediation methods and standards of cleanliness.

A review of community involvement by the GAO at eight Superfund sites found that the EPA did abide by statutory mandates by, for example, developing community relations plans, establishing information repositories, issuing public notices of meetings, and incorporating community comments and EPA responses to them in the final cleanup plan (US GAO 1994b). Although the letter of the law was followed, the report concluded that the spirit was often lost. Among other complaints, residents felt that the EPA was late in soliciting community input (after a site was placed on the NPL), involved too few community members, did not give adequate attention to residents' health concerns, and did not effectively explain technical information.

There is substantial evidence in the literature on environmental risk communication of the difficulties involved in winning public trust and achieving consensus with respect to cleanup standards and methods (Kasperson and Stallen (eds) 1991). However, there is also agreement among legislators, environmentalists and community members that the effort must be made and that changes in the programme's structure would help to provide citizens with more meaningful input (e.g., soliciting community opinions earlier in the process). Some of these suggestions have been articulated in the Superfund Reauthorization Act proposal discussed at the end of this chapter.

FINANCING THE FUND

CERCLA combines a regulatory process and imposition of requirements for cleanup of past pollution on potentially responsible parties (PRPs) with the special Fund, which is used for site mitigations conducted by the EPA itself, or to cover 'orphan shares' of cleanup costs for PRPs that are no longer financially viable or have ceased to exist. The regulatory functions are financed by the general revenues of the federal government, but the Fund draws its revenues from a special tax. It is this special taxing capacity that requires successive reauthorization under provisions requiring that any such earmarked revenue generators lapse after some period of time. Thus the debates over financing CERCLA have focused on how the Fund revenues are to be generated.

Numerous alternative means of financing the Fund were considered when CERCLA was first passed and during later reauthorizations. Taxing hazardous waste directly seemed unwieldy, as it would have entailed fee collection from over 200,000 waste generators. Moreover, assessing generators on the basis of the volume of waste they reported would increase their incentives to falsify RCRA records and to illegally dispose of wastes (Landy et al. 1994). A fund financed by a fee on petroleum products and chemical feedstocks was deemed more workable and was adopted.

Taxing industry fostered the public perception that the polluter pays principle was being implemented — that those who created the pollution were being compelled to pay for cleanup. Indeed, the fee targeted industries responsible for the lion's share of all hazardous waste (US CBO 1985). In 1986, SARA expanded the five-year Fund from the original $1.6 to $8.5 billion and adjusted the tax burden somewhat to include general revenues. However, the principle of taxing some subset of principal toxics generators has been consistently retained.

In actuality, consumers pay through higher product prices, but the costs are hidden and are therefore palatable. It is difficult to determine whether or not reliance on higher general revenues would produce more equitable financing than targeted producer taxes that are passed to consumers. However, the argument that a consumption tax may be more progressive than one on income, or at least more equitable relative to demand for goods derived from chemical feedstocks, appears to have prevailed.

Another issue that has arisen in public debate on the Fund involves resource recovery problems. Initially, the Fund was intended to be partially self-replenishing. That is, when the EPA spent Fund revenues to clean NPL sites, it was to recover costs from existing, financially viable PRPs. In practice, it has not entirely succeeded in this regard: the Agency has reached agreements with PRPs to recover only $1.2 billion of the $8.7 billion it spent through 1993. It must be noted, however, that, in addition to recovering costs incurred by the EPA, the Agency also supplements cleanup funding by compelling PRPs to clean sites under the principle of strict, joint and several liability, to which we turn next. Here, the EPA's efforts have been substantially more successful. As of the end of fiscal year 1993, the Agency had entered into settlements with responsible parties for over eight billion dollars in remediation costs (US GAO 1994c).

LIABILITY PROVISIONS UNDER CERCLA: EQUITY ISSUES

The Superfund is unique among industrial nations' approaches to financing remediation in its efforts to place the burden on those private and sub-national governmental units that generated the contaminants. The equity and efficiency questions raised by the approach have proved very contentious and remain subject to question. We turn next to a discussion of the principles and impacts of assignment of strict, joint and several liability.

Strict liability means that responsibility for cleanup is assigned without the need to prove negligence or fault. Its rationale is the rigid polluter pays principle: if contamination resulted, those associated with it, regardless of fault, should pay for mitigation and damage. The effect is that parties who were law-abiding at the time the pollution was created can, retroactively, be held liable for damages deemed illegal under later legislation. This differs from most injury claims in the US that ordinarily can be brought only under legal requirements prevailing at the time the actions were taken and require specific assignment of fault (Church and Nakamura 1993).

Joint and several liability means any one party can be assigned the full responsibility for harm caused by several parties. This liability principle was not explicitly legislated under either CERCLA or SARA, but has emerged as the result of a number of judicial decisions. (For discussion of these court cases, see Fogleman 1992, Forte 1991,

Segerson 1992, and Toulme and Cloud 1991.) The rationale for imposition of the liability scheme is threefold (Church and Nakamura 1993):

1. Promoting settlements: it encourages PRPs to negotiate a cleanup and avoid costly litigation. It is intended to bring recalcitrant parties forward to avoid law suits.

2. Pursuit of 'deep-pockets': it provides leverage for the EPA to force firms with substantial resources to pay for cleanups. In principle, this approach was to make it easier to replenish the Superfund — or, better yet, to shift the full cost burden on to those PRPs with the financial capacity to conduct the mitigations themselves.

3. The incentive effects: it has a positive influence on the future behaviour of those handling toxics. Knowing that they may be held responsible for *all* costs associated with contamination, all contributing parties would exercise discretion and, in effect, take over the enforcement of CERCLA and RCRA standards on the operators of the disposal sites they utilized.

In practice, these provisions have meant that the entire chain of owners of a property, and, as we shall discuss, potentially their advisors and other investors, could each individually be held liable for any and all consequences of contamination on the site, whether or not the damage occurred while any one party held title. The magnitude of their pollution contribution thus becomes irrelevant to their potential cost share. Acceptance of full liability by one party cannot absolve others of potential liability in the event that the costs of cleanup or damage mitigation exceed the assets of the individual or organization accepting responsibility.

The liability provisions facilitate EPA action, permitting it to pursue one party for the full cost of a contamination. Common law, however, permits that party to then sue other contributors to the pollution for their 'shares' based on their prior disposal actions. The result has been massive civil litigation among the PRPs themselves (Acton and Beider 1992). Such legal actions have delayed voluntary cleanups, engendered extensive unanticipated private litigation costs, and exposed parties that were not intended targets of CERCLA enforcement to liability.

The common liability of a number of parties for many of the NPL sites and properties otherwise regulated under CERCLA has produced a range of equity problems and affected the operation of capital markets. In aggregate, the result has been massive transaction costs, largely for litigation, and borne predominantly, but not exclusively, by the PRPs. A RAND study of eighteen NPL sites estimated that transaction costs constituted 32 per cent of cleanup project costs and that 65 per cent of these costs stemmed from legal work (Dixon et al. 1993).

The major equity considerations that have been raised, exclusive of retroactive liability for actions that were legal when taken include impacts on *de minimis* contributors and prospective or future liability. Both have some impact on capital flows, but this issue is more directly affected by lender liability provisions, which also affect the liability of a whole array of public sector organizations not directly involved in waste generation or disposal. We first examine *de minimis* contributors and future liability, then address lenders' concerns, and finally turn to impacts on other private actors and public sector operating agencies. This sequence moves us from the more particular to very broad equity and efficiency considerations.

De Minimis Contributors

One aspect of CERCLA that received especially harsh criticism is the liability exposure of *de micromis* and *de minimis* contributors (that is, those who have contributed minuscule and small amounts of pollution to a site), or that of those who held title to a property for a very short period, or who had only minimal power to influence management of a facility generating or disposing of toxics. These parties have rarely been the subject of EPA enforcement. However, they are drawn into litigation through contribution suits brought by other parties. To address this issue, SARA authorized expedited settlements with *de minimis* parties, which provided small volume waste contributors an early opportunity to fully resolve their liability for a site. However, these procedures have been underutilized for various reasons including the costs to EPA of collecting data necessary for making the settlements (US GAO 1994d).

The complications and costs potentially presented by such parties are overwhelming: the number of possible lawsuits among PRPs is, theoretically, the factorial of their number, when all possible combinations are considered. Thus a site with five PRPs could produce

5×4×3×2, or 120 different lawsuits. If the site had five substantial contributors and another five minimally involved parties, however, the possible suits would grow to ten factorial, or 3,628,800! Clearly, the theoretically possible complications have not arisen in practice, but the relative complexity of arriving at privately negotiated or litigated cost allocations, or of EPA-imposed settlements rises spectacularly with the number of PRPs involved. The Wichita, Kansas, contamination case, which we describe later in this chapter, involved over 500 PRPs; the voluntary agreement arrived at there was clearly influenced by fears of the costs of litigation.

Future (Prospective) Liability

One of the peculiarities of the liability for contamination under CERCLA is that responsible parties are not released from liability upon settlement. Parties remain vulnerable to future claims that may arise many years later as the result of improved contamination detection or new findings on damage done by past pollution. CERCLA specifically requires the EPA to include in most settlements a reopener provision for future liability based on unknown site conditions. Thus, no matter what the current answer to the 'how clean is clean?' question may be, the issue can be reopened in the future. The law states, in effect, that no cleanup is ever completed in terms of the liability of responsible parties.

This provision is the most problematic from the point of view of would-be site developers, in terms of its impact on the expected returns on investment in reclamation and reuse of contaminated sites. As suggested in Chapter 1, this prospective liability leaves future costs open, imposes possible burdens on developers for continued site monitoring, and otherwise reduces the economic value of even those sites that have successfully been cleaned in conformance with *current* CERCLA requirements.

Problems of Lender Liability

Owners of property or operators of facilities are liable under CERCLA. The law, however, explicitly excludes from the category of owner or operator any party who, without participating in its management, holds 'indicia of ownership' such as a share in title or foreclosure right for the primary purpose of protecting a security interest in the property. However, in court cases such as US v. Maryland Bank & Trust

Company in 1986, the statute has been construed narrowly, indicating that a mortgagee may lose its exemption from liability when it forecloses, thus becoming the owner. Banks and other lenders, therefore, have adopted the policy of conducting site contamination assessments before they foreclose and are sometimes walking away from their collateral to avoid environmental liability (Seidman 1991).

A commercial lender with a borrower in financial difficulties has an option to act to avoid foreclosure, which is preferred from the perspective of maintenance of economic activity: involvement with its borrower's business and provision of management advice in order to preserve repayment abilities. Such practices, however, appear threatened by the spectre of CERCLA liability. Another case, US v. Fleet Factors Corporation (1990), produced a legal finding that a lender may be held liable because it had the *capacity* to influence the management of hazardous waste materials, even though it may not be so involved (Toulme and Cloud 1991). This decision sent shock waves through the lending community. One environmental risk specialist with a major commercial credit corporation who had previously served in the EPA concluded that, in the light of the judicial findings, lenders 'have no incentive to enter into transactions that involve the material risk of environmental liability' (Campbell 1992:56).

Given continued lending for brownfield redevelopment, this statement is clearly an exaggeration, but it highlights the financiers' dilemma: they can attempt to encourage their borrowers to be more responsible with respect to handling hazardous materials and risk being found to have participated in managing the firms, thus exposing themselves to liability. Or, they can ignore the borrowers' activities and risk non-payment of the loan if CERCLA-imposed remedies, costs and penalties cause their client to become insolvent. The loss associated with loan defaults is known; that resulting from a share of CERCLA liability is uncertain (and could exceed the value of the loan itself). The legal findings, then, result in incentives for lenders *not* to act as responsible corporate citizens, assisting to minimize pollution, but to sit idly by, even when they know that some of their clients are contaminating their properties (Forte 1991).

The irony of this situation is that the EPA and the lenders actually have a common interest in restraining inappropriate handling of toxics. Admittedly, the EPA is concerned with contamination while the lenders worry about the financial consequences of their clients running foul of CERCLA, RCRA and related regulations. Nonetheless, both gain if violations are avoided, and the lenders are in a position to monitor their borrowers far more closely than can a resource-constrained federal agency tracking tens of thousands of sites. However, as Edward Kelley (1991) of the Federal Reserve System notes, lenders do not generally have the technical expertise to police the environmental aspects of a borrower's operations.

To help ease the lenders' predicament following the Fleet Factors decision, the EPA issued a new rule on lender liability in 1992 (US EPA 1992). The regulatory intentions were, first, to allow lenders to work with their borrowers without necessarily incurring liability (for example, the rule provides that a lender is not an owner or operator unless it actually exercises decision-making control). Second, the rule sought to preserve the collection remedy of foreclosure while ensuring that the benefits of a taxpayer-financed cleanup did not produce windfall gains for private lenders (for example, following foreclosure, the lender must attempt to divest of the facility in an expeditious manner). The rule, however, failed to resolve a number of ambiguities, especially as regards possible PRP suits against lenders, thus failing to alleviate their fears in the light of the proactive interpretations of the courts in extending liability (Schnapf 1992; Witkin 1992). More importantly, the regulations dealt solely with lender liability under Superfund, not under other federal or state environmental laws and many state statutes.

It has been suggested that the interests of the EPA and lenders would correspond even if federal law and regulations completely protected lenders from any environmental liabilities. We find these arguments extremely persuasive. Harris Simmons (1991), representing the American Bankers Association at Senate hearings on CERCLA lender liability, notes that no federal liability protection rule will protect lenders against a decline in the value of their collateral due to environmental impairment. Lenders still will have a tremendous incentive to continue the practice of conducting environmental audits on projects they finance because, if the properties are polluted, the lenders still stand to lose the entire value of their loans: first, the security value of a given property is reduced by pollution; second, the borrower is given an incentive to declare bankruptcy and walk away from the property to avoid contamination liabilities; and, third, if the borrower tries to correct the damage, he or she will incur cleanup costs that would impair the ability to service the debt. Thus, even in the absence of any lender liability, ample incentives for lenders to continue serving as environmental 'detectives' would remain.

Other Unanticipated CERCLA Liability Effects

Beyond those parties discussed above, other demonstrably innocent parties are legally exposed to CERCLA liability. First, municipalities often utilize (and may manage) landfills which also receive some industrial toxins. CERCLA does not specifically address the generators or transporters of municipal solid waste and, thus, does not exempt them from potential liability. The EPA does not generally pursue compensation from municipalities for their involvement in landfill operations. However, when charged with responsibility for contamination at landfills by the EPA, major polluters in turn file suit against the municipalities, thus placing municipal solvency at risk (Glickman and Judy 1993; Tomsho 1991).

An additional set of inculpable but legally exposed parties include those who inherit properties they have never managed and a variety of corporate and individual agents who hold property in a fiduciary capacity (e.g., executors and trustees of estates and trusts). In many cases, these parties will have had virtually no knowledge of the nature of the property in question until they inherit it or their fiduciary duties become operative. However, they may be found *personally* liable as 'owners' when response costs are incurred for contamination on the land involved.

Returning to the federal government, we have already noted that such operating agencies as the Department of Defense and Department of Energy face facility cleanup costs that may far exceed CERCLA costs. However, the Superfund programme also poses potential liability risks to a series of other federal agencies. These arise in three forms. First, because of the volume of mining conducted on federal lands; second, because any contamination resulting from response actions taken under CERCLA is, itself, subject to the law's regulatory requirements; and third, because the federal government is a major financial and fiduciary institution in its own right, with a number of agencies facing the same liability risks as private financiers.

Mining activities create hazardous substances, including deep mine tailings and surface mine overburden. According to the US General Mining Law of 1872, a person with a mining claim controlled use of the property, so the federal government was not responsible for environmental problems that arose. However, the 1976 Federal Land Policy and Management Act conferred an ability to 'manage' mining activities permitted on federally-held lands on the Bureau of Land Management, National Parks Service and other federal landholding agencies. As a consequence of this law, all now face potential CERCLA liability claims from the mining corporations that used their lands.

Beyond the EPA itself, a number of other federal agencies may engage in emergency cleanups or other responses to contamination under CERCLA authority. These organizations, among them the landholding agencies which might pursue cleanups of mines, the US Coast Guard and the US Army Corps of Engineers, all then face liability risks comparable to those of private contractors under contract to the EPA or PRPs to conduct cleanups of contaminated sites.

The full panoply of federal agencies and quasi-public agencies engaged in financial activities is too lengthy to describe here. Entities such as the Farm Credit Administration and Small Business Administration face the same problems with their borrowers as the private financiers we discussed above. Somewhat different issues arise for the institutions created to protect the deposits of private parties who have entrusted their funds to financial institutions that fail, such as the Federal Deposit Insurance Corporation (FDIC), the Federal Savings and Loan Insurance Corporation and the agency created to cope with the savings and loan débâcle of the 1980s, the Resolution Trust Corporation (RTC). None of these agencies hold properties to manage them or earn income from them, but rather to dispose of them on the market in order to raise the capital to repay depositors in the failed institutions. They thus face problems akin to municipalities taking properties for failures to pay real estate taxes or private financiers foreclosing on loans.

The RTC alone has been disposing a portfolio of over 36,000 properties as part of providing a bailout of failed savings and loan institutions that amounts to over $500 billion. Its potential CERCLA liability could swamp even its massive budget if its experience parallels the FDIC finding that cleanup costs on a sample of 117 contaminated sites they held amounted to roughly 1.5 times their estimated market value (Seidman 1991).

THE IMPACT OF CERCLA ON BROWNFIELD REDEVELOPMENT

That CERCLA liability has significantly retarded efforts to renovate brownfield lands and buildings through its impacts on perceived real estate investment returns is a belief widely promoted by policy analysts and practitioners (Bartsch and Collaton 1994; Glaser 1994) and promulgated in congressional hearings testimony (US House 1989, 1991; US Senate 1991, 1993). Uncertainty about current contamination and the cost of required cleanup or mitigation is thought to deter new investment. This effect has been linked to broader problems of urban regeneration, with the argument that 'brownlining' results in capital starvation of areas known to have — or suspected of having — contamination. One major commercial lender stated explicitly that 'it was more prudent for the company to draw a large circle' around an area suffering from contamination than to risk any investment there (Pollard 1992:67).

Cities and urban areas contain a variety of brownfield areas other than heavy industrial districts which may prove problematic under CERCLA. Possible difficulties exist in residential areas that may contain asbestos or be located over former landfills, gas stations and other businesses with underground storage tanks such as auto dealerships and fleet operators, auto repair shops, dry cleaners, tool and die shops, wood preserving facilities, scrap yards, utilities, and bottling and canning facilities.

Notwithstanding the presence of all such potential contaminators and polluted sites, the presumption that CERCLA actually substantially inhibits regeneration and the flow of capital into urban areas has not been verified empirically. The capital access problems of urban areas may be due to the existence of better investment opportunities in other locations, independent of the presence or absence of contamination. The general tendency towards urban sprawl and suburban development at the expense of central cities is a US pattern that long predates the passage of CERCLA (Birch 1970).

We examine here the apparent effects of CERCLA on general access to capital, then look at transaction costs facing lenders in the light of CERCLA requirements, and conclude with some findings on the Superfund's impact on expected returns on investments in brownfields. However, we must recognize that, in the light of the established US urban abandonment pattern, these CERCLA effects may not have materially diverted capital from brownfield sites in settings rejected by investors for non-environmental reasons.

CERCLA Impacts on Access to Capital

In 1986, SARA added what has come to be known as the 'innocent landowner defense' which exempts from liability any owner who had no reason to know about existing contamination. To make this claim, a new owner must conduct appropriate inquiries into the previous uses of the property. Consequently, lending institutions now routinely require environmental assessments for transactions involving sites previously used for industrial and many commercial purposes (Forte 1991; Kelley 1991; Segerson 1992). Such assessments raise the cost of processing loans, and thus the cost of capital to would-be borrowers.

The assessments have the secondary benefit of uncovering pollution that needs to be treated. As we noted, CERCLA has successfully turned the financial sector into an environmental detective agency. The price of that transformation appears to be a highly undesirable impact on urban regeneration: reduced availability of capital at any price.

A 1991 survey of its members by the Independent Bankers Association of America found, for example, that 86 per cent of respondents stated that environmental concerns affected their lending policies over the prior five years. Seventy-three per cent reported that there were loans they would not write due to environmental concerns. However, only 21 per cent claimed a loss or default due to environmental impairment (Witkin 1992). William Seidman, Chairman of the FDIC, reported in 1991 that no bank or thrift institution had suffered a debilitating loss or failed as the result of Superfund liability, noting that this finding arose in large part *because* they avoided environmentally risky loans.

Two distinct questions of immediate causality arise from this apparent contraction in the availability of capital for urban regeneration. The first is the role of loan transaction costs in restricting small borrower access to capital. The second is the effect of contamination on the potential losses to lenders if they abandon their investments, on borrower capacity to service debts and on the collateral value. The former is a matter of accounting and processing realities; the latter appears to be a matter of belief and perception, not borne out by empirical evidence.

Lender Transaction Costs

Transaction costs for loan underwriting alone impose burdens on all development and redevelopment projects. To the extent that new monitoring and assessment of environmental risk exposures is routinized in loan assessments, which has been a consequence of CERCLA, all loans cost lenders more to make. They, in turn, pass those costs on to borrowers in higher interest rates. The higher cost of capital inevitably slows the adaptation of land use to changing economic conditions. Moreover, since transaction costs are only partially sensitive to the scale of the transaction, such cost burdens may raise the minimum threshold of loans an institution would be willing to process.

A basic environmental assessment for a small commercial site offered as loan collateral might cost $500–$2,500, which amounts to one to five per cent of the value of a $50,000 loan. Those costs would have to be paid, one way or another, by the borrower, thus reducing his or her effective loan proceeds. The assessment would not guarantee protection for the lender but, in reducing the loan value to the borrower, weaken his or her ability to service the debt. Many institutions may decide that the risks involved in such transactions, producing profits of only $1,000 annually if all went well, are simply not worth accepting.

A rising threshold for new loans will affect urban economies in two ways. First, it may threaten the viability or expansion of new, small and growing businesses which need regular small injections of new capital but are not capable of taking on large amounts of debt at any one time. A rising minimum loan level forces them to operate without new capital until they get large enough to service a bigger loan, thus increasing the risk that they will fail before they can expand to that level. Second, it has serious consequences for the reclamation and redevelopment of small parcels of potentially contaminated land. The small sums associated with reclamation of small commercial sites are not likely to be available, given that development loans present more risk of losses associated with CERCLA liabilities than even the commercial loans we have examined.

Perceived Potential Losses Attributable to Contamination

Because of environmental liability concerns, many lenders may forgo brownfields investments on a wholesale basis. In doing so, the distinct possibility arises that they are missing profitable investment opportunities, that is, their anticipated losses associated with cleanup liabilities may be systematically exaggerated. One reason for this exaggeration may stem from the fact that expectations about costs of contamination treatments appear to derive from data on NPL sites. Since NPL sites constitute only about 4 per cent of the sites on the EPA's list of possible sites which, in turn, are under 10 per cent of the estimated 400,000 or more contaminated sites in the US, such generalizations are unwarranted. Yet the NPL data appear to influence financiers' decisions on projects that involve potentially contaminated properties. In part, this may be due to the fact that no national data exist on remediation costs of non-NPL properties.

Edward Kelley, for example, representing the Federal Reserve System at a Senate hearing, argued that CERCLA reduces the willingness of lenders to provide credit to businesses by noting that, 'With the average projected cost of remedying contamination at sites on the National Priority List climbing to over 25 million dollars, liability in CERCLA cases may far exceed the amount of the lender's original loan' (1991:101). Similarly, Representative John LaFalce of New York opened hearings on the negative impacts of CERCLA on small businesses by remarking that, 'you may be saddled with the entire cost of cleanup....That cost regularly runs over $1 million per site, and sometimes tops $10 million' (1989:1). In testimony presented in those hearings, however, the Small Business Administration (SBA) presented data on its losses as a lender to small businesses (the very type of enterprises expected to present the greatest risk to financial institutions) which contradicted Representative LaFalce's assertion.

Surveying their regional offices, the SBA found 140 cases involving contamination problems and agency losses (US SBA 1989). The data are incomplete but indicative:

1. the 140 cases cover roughly eight years' experience under CERCLA for an agency making thousands of loans annually, and include many loans made before the bill was passed, so *the incidence of problem loans is low*;

2. only one of the cases was an NPL site, so *experience with NPL cleanups is not a valid basis for cost expectations*; and,

3. projected losses due to cleanups required on properties owned by SBA averaged under $300,000, and losses due to abandoning properties with excessive cleanup burdens were under $550,000 per site, so *claims of minimum losses of $1 million or more are not borne out*.

Since SBA is a government agency, not a for-profit lender, its efforts at due diligence, assaying risks due to possible past or current contamination of property, would be expected to be lower than those of private financial institutions. The efforts of the latter should enable them to avoid even the relatively low losses recorded by the SBA. However, even if we extrapolate the experience of this one lending non-profit institution to the financial sector as a whole, we would have to conclude that fears of major losses due to CERCLA liabilities may well be exaggerated.

Further evidence is manifest in data on the property values of contaminated sites and properties near them. The common perception and fear, which we identified in Chapter 1, is that the properties affected with contamination become permanently stigmatized, suffering significant (and perhaps unwarranted) decreases in their value. The evidence on permanent stigma is mixed, however, and the evidence suggests that, even at NPL sites, prudent management of property portfolios could avoid any financial losses due to depressed collateral values.

Admittedly, nearby property value declines are most pronounced when a site is added to the NPL — a newsworthy event. However, the decline may disappear once the site is remediated (Hird 1994). So, if current owners purchased the land at a discount after a site was discovered, site remediation could provide a windfall profit. Owners who sold on learning of the hazard would suffer losses. Financial institutions potentially acquiring the properties due to loan defaults might suffer some losses due to delays, but, by holding properties until mitigations were completed, they could benefit from the cleanups if they themselves were not liable for the decontaminations.

In general, earlier studies failed to produce evidence of a significant property value loss due to stigmatization, even for lands adjacent to facilities such as waste disposal sites (Zeiss and Atwater 1989). More recent evidence suggests a direct connection between contamination and value. Page and Rabinowitz (1994) found that on-site groundwater contamination depressed commercial and industrial property values; their results indicated effects ranging from 10 to 50 per cent in their six case studies. Looking specifically at sites with leaking underground tanks, Simons and Sementelli (1994) found that properties with such problems were less than 50 per cent as liquid as similar uncontaminated sites. Their findings are consistent with the general observation that sales on contaminated properties are generally slower than those on clean sites, and prices suffer deep discounts. Similar effects were recorded by Simons and Sementelli for unpolluted residential sites close to leaking underground tanks in Cleveland, which they found suffered an average 17 per cent decrease in property values (Simons and Sementelli 1994). Page and Rabinowitz (1994), by contrast, found no effect on residential properties that actually suffered groundwater contamination.

Mundy (1992b) concluded that the value assigned to cleaned properties was lower than the sum of the value of the sites when contaminated plus the cost of their cleanup, claiming the presence of property stigmatization by prior contamination. Peiser and Taylor (1994) found the exact opposite: no overestimation of risk associated with cleanups and a smoothly operating and efficient property market. Their conclusion, that knowledgeable developers can capitalize on the ignorance of other parties in real estate markets, suggests that many individual market participants are exhibiting excessive levels of risk-aversive behaviour. If the market does operate efficiently, then those responding to stigma stand to lose, relative to better informed investors.

None of this evidence, however, appears to justify the repeated declarations in congressional testimony by financiers and their clients that collateral value is so depressed by contamination that lenders stand to lose their investments. Even the highest losses reported — 50 per cent — would not be damaging if loans were limited in value to that fraction of the real estate collateral offered. Since US financial institutions currently lend no more than 80 per cent of appraised value for owner-occupied residential property without some form of lending insurance, and generally offer only 65 to 70 per cent for rental residential units, a 50 per cent standard is not unreasonable for non-residential lending.

We can thus conclude that lenders' claims of excessive and inappropriate loss exposures may be justifiable, but only in terms of their possible institutional liability for cleanups themselves, not on the basis of loss of collateral in most instances. It is further clear that EPA itself has been attempting to protect innocent lenders who are not involved in contaminating operations from such liability. Whatever inappropriate risks may be present for lenders appear to derive from the attempts of landowners and occupiers who, when designated PRPs for contamination, search for other parties on to whom they can shift some of the cleanup cost burdens.

Whether or not they are valid and justified, investors' cost expectations and fears lead them to reduce their valuations of potential returns. To the extent that such assessments constitute *undervaluations*, many projects that might be economically viable may not be undertaken and both individual municipalities and the economy as a whole may be deprived of possible welfare gains due to the uncertainty and potential liability imposed by the Superfund legislation. Whether or not CERCLA actually depresses property investment returns, and regardless of its real impact on willingness to invest in brownfield sites, the individual US states have responded to the perceived problems with Superfund liability with a variety of special subsidies and programmes to stimulate reclamation and regeneration of such properties.

STATE LEGISLATION TO 'SOLVE' THE REDEVELOPMENT PROBLEM

Superfund legislation and the regulatory efforts of the Environmental Protection Agency to implement and enforce the laws mandating cleanups and imposing liability for hazards have treated the problem of polluted land as a matter of *environmental* policy. The individual states (and many of the cities and counties) have all created their own environmental agencies in order to address an array of pollution problems. The state entities, in particular, have taken over supervision of many contaminated sites not on the NPL. Those that are approved by the EPA may also be delegated the authority (and financial resources) to conduct NPL site cleanups.

The states also take a different legislative approach to abandoned or underutilized brownfield sites: various efforts to assure that contaminated lands get redeveloped have generally approached the reclamation and mitigation problem as a matter of *economic development* policy. There is thus a tension between the two types of legislation and their intents.

State and local governments are attempting to address CERCLA-related problems by establishing buyer protection programmes (with states certifying cleanliness and agreeing not to sue for further detoxification), offering tax abatements and financial assistance for cleanups, and the like. We examine state laws here under three broad categories.

Property Transfer Laws and Covenants Not to Sue

Several states have buyer protection programmes in the form of property-transfer laws. Some combine state assurances of limited liability with tax abatements and other subsidies to new investment on polluted lands. State legislation is not uniform, and there is a broad range in the speed with which the states have responded to the investment constraints inherent in the Superfund laws.

New Jersey, a small state, burdened with more NPL sites than any other, passed its Environmental Cleanup Responsibility Act in 1983. Under the law, sellers *must* conduct environmental assessments and prepare cleanup plans that meet state standards before any transfer of polluted property can take place. (Connecticut, Illinois, and California laws, by contrast, only modify *caveat emptor*, simply requiring pollution disclosure prior to sales.) The state then certifies cleanliness to buyers and provides a covenant not to sue over contamination for landowners proposing to redevelop or reuse a previously contaminated site, reducing the uncertainty they would otherwise face. Illinois has had a similar law since 1988. However, the state is still having problems finding owners, and many holders of polluted lands find it cheaper to sit on their properties rather than clean them up.

Section 14A of Michigan's 1990 Environmental Response Act permits the state to grant covenants not to sue provided that a redevelopment project or land transfer meets a string of public interest conditions covering both environmental and economic benefits. The action must: (a) yield new economic resources, (b) expedite environmental response and mitigation action, (c) not worsen the release of contaminants or pose health risks, and (d) exhibit economic development potential. The Minnesota Land Recycling Act of 1992 also provides a statutory covenant not to sue that covers developers, their lenders and possible successors, provided all are non-responsible parties, after the state approves voluntary response action plans and certifies the cleanup but does not impose public benefit requirements for state approvals.

Indiana only recently (July 1993) passed a law empowering its state Department of Environmental Management to work directly with landowners on site mitigation and certify cleanliness. Ohio has also joined the states offering covenants not to sue, but has added the 'sweetener' of a five-year tax abatement on property value increases due to mitigation — and could permit the Ohio Environmental Protection Agency Director to grant variances from state cleanup guidelines.

These are but a few examples of state efforts to limit the impediments associated with the broad liability imposed under Superfund. However, federal law takes precedence over state legislation. Thus any tendency for the states to provide opportunities for less stringent mitigation in order to promote redevelopment and reuse may backfire: the covenant not to sue offers protection from state action, but not from federal enforcement. Moreover, for sites that are close to state lines, leaching or migration of in-ground pollutants may permit adjacent political jurisdictions to bring actions for damages under federal law. The protection from liability offered by such provisions, then, depends critically on the enforcement capacity of the EPA and the extent to which the federal agency relies on the agencies established by the states to act as enforcement agents (since the state offices are more likely to abide by their jurisdictions' commitments not to prosecute or sue).

Financial Assistance Programmes

An alternative to legislation that reduces apparent liability exposure is provision of public sector funds to reduce the costs associated with investments in reuse projects. This approach has been promulgated in a number of states. (Federal legislation to provide some cost coverage has been introduced repeatedly, but not enacted, in recent sessions of Congress.) State financial assistance has taken two basic forms.

First, there is support for site assessment and contamination detection costs. Connecticut explicitly provides funds for brownfield site evaluation and remediation planning under its 1993 Urban Sites Remedial Action Program. Michigan's Site Reclamation Grant and Loan Program contains a similarly earmarked pool of funds. However, virtually all the states have economic development funds that, one way or another, can provide funds for feasibility studies and redevelopment plans designed to attract new businesses and jobs. These funds have been used to varying degrees to subsidize the transaction costs associated with reuse of brownfield sites, depending on the priority placed by the states on regeneration relative to greenfield development.

The second form of financial assistance provided is addressed directly at the returns on investment if a project is undertaken on a brownfield site. Support can take the form of either grants or loans, with the former imposing more costs on the states and permitting windfall gains to developers. Pennsylvania's Industrial Community Action Program, for example, offers grants to bring blighted industrial sites into productive reuse for new *industrial* activity. The subsidy is thus available only for a particular form of job generation. Another set-aside pool of funds under the Michigan reclamation programme provides funding to local governments for site remediation based on environmental factors as well as economic development potential.

Purposes include reducing pressure on green spaces by recycling sites in areas having existing infrastructure, supporting projects which provide jobs and increase the tax base, and facilitating the use of a mixture of funding and private investment. New Jersey's Hazardous Discharge Site Remediation Fund can be used only for loans to private parties to fund remediation required under the state's environmental laws (but, out of another pocket, provides grants to municipalities for assessment and cleanup).

An alternative to direct subsidies for projects is, of course, provision of tax abatements on appreciated property values. This tool simply increases the net earnings on any cash flow from a property investment. Since it does not require any up-front cash outlays, it is easy to implement. However, the invisibility of such 'tax expenditures' (revenue forgone) has led to their overuse by the states for economic development. Therefore, it is difficult to provide the extraordinary subsidy needed to make potentially or actually polluted brownfield site projects preferable investments to pristine greenfield alternatives. Moreover, there is little reason to expect subsidies to redevelopment of brownfield sites to be particularly successful, especially since many of the programmes that provide public funds for site mitigation also provide public support for other development efforts on unpolluted land (Barnekov et al. 1989; Bingham and Mier (eds) 1993).

Permitting Variation in Standards of Cleanliness Depending on Future Use

An alternative to provision of subsidies for either site investigations or actual cleanup operations is, in principle, available, to the individual states, albeit the principle is explicitly rejected in the federal legislative approach. Standards for site mitigation could be varied, depending on the zoning and intended future use of polluted lands. New Jersey, with its exceptional concentration of NPL sites and other contaminated properties, has been the innovator in this regard, and has promulgated different cleanup standards for residential and nonresidential land. It is not alone, however: the flexibility mandated in Minnesota and Ohio legislation, for example, effectively implements similar distinctions.

At issue here is the *necessary* level of remediation to attain the health and environmental safety goals undergirding the legislation requiring cleanups. The risks from contamination are: (a) dangers to the humans using the land for whatever purpose it is to be redeveloped, and (b) dangers associated with the spread of pollution from the site to other areas (through groundwater or movement through the soil of insoluble pollutants). When off-site migration of contamination is not a threat, the future use of the property is logically the key determinant of the level of cleanup required to protect human health. The existing or future zoning or other land-use controls on a site could thus provide the primary basis for variation in mitigation standards.

CERCLA itself does not make any such distinctions (and, as we have noted, the law and its implementing regulations have not, to this day, clearly defined a standard of cleanliness). If state law permits lower levels of mitigation for reclamation of sites intended for, say, industrial uses than CERCLA minima established by the EPA, then either the developer or the state would be liable for federally-ordered additional mitigation. On the other hand, variable standards permit states that want to respond to exceptional community concerns about pollution to set standards *above* those promulgated by the EPA in order to attract, for example, higher-income residential uses for centrally located urban brownfield sites.

The US practice of hierarchical zoning, under which any land use up to a specified level of intensity is permitted on a site, creates some problems for variable standards: in principle, this form of zoning permits construction of single family homes on land zoned for heavy industry. While some exclusionary zoning, dictating the use of sites that is permitted, is present in a number of locations (New York City provides one example), American efforts to maximize individual choice has tended to result in policy that sets limits rather than dictating uses.

Application of exclusionary zoning and variable standards for cleanliness to brownfield sites could significantly reduce the costs of reuse for commercial and industrial purposes while not generating a threat that housing might be placed on contaminated land. Thus the variable standards some states are promoting would provide impacts on the costs of new economic development projects directed at job creation comparable to their investment subsidy programmes, only without the expenditure of public sector revenues.

RECENT CERCLA REAUTHORIZATION AND CHANGE PROPOSALS

Recent congressional hearings on CERCLA number in the dozens. Reports by federal 'watchdog' agencies such as the Government Accounting Office, Congressional Budget Office and Office of

Technology Assessment may be counted in scores. The EPA itself has generated literally hundreds of evaluative and planning documents, and discussion papers reviewing CERCLA problems and potential resolution approaches. The best means of summarizing all these inputs to the current debate over directions for contaminated land policy in the United States is simply to outline the major provisions of the Clinton Administration's 1994 proposal to reform the Superfund programme.

That proposal, submitted to the Senate and House of Representatives in February 1994, was itself the subject of a series of hearings, was amended, and was not brought to a vote. The suggested changes, however, serve to summarize the major elements of concern with, and prominent proposals for changes in, the programme as it has operated to date.

The Administration's proposal to reform Superfund included

1. Provision for greater involvement of communities near NPL sites. New community-based organizations would be formed to promote public participation throughout the Superfund process, beginning with the site assessment phase, and would be provided with resources to access needed technical expertise. The community's preference with respect to future use of the land would be considered in the development of remedial alternatives.

2. Implementation of Environmental Justice amendments to assist disadvantaged communities facing multiple sources of pollution. The NPL Hazards Rating System would be modified to take into account the presence of multiple-risk sources within any one geographic area, which it does not now do.

3. Reduction of the current potential for duplication of federal and state government responsibility at NPL sites. States would be offered the choice of assuming authority for sites and given access to Fund resources, provided they employed mitigation standards at least as stringent as those applied by the EPA.

4. Limitation of 'innocent' purchaser and lender liability to encourage brownfields redevelopment and facilitate voluntary cleanups orchestrated by states at low- and medium-risk sites. Purchasers' liability would be limited if they acquired the property subsequent to disposal of hazardous substances,

conducted a site audit and, in the case of property for residential or non-commercial use, a site inspection and title search revealed no basis for further investigation. If federal funds were used for cleanups, the government could place a lien on the property, based on the net difference between the value of the property prior to and following the response action. The lien would continue until it was satisfied or all response costs were recovered.

5. Modification of the Superfund liability scheme to increase fairness and efficiency in several areas:
    a. exemptions would be provided for *de micromis* parties who have contributed minuscule amounts of contamination at a site;
    b. utilization of expedited settlements would be increased for *de minimis* parties or those contributing minimal amounts of toxics (authorized under SARA) by easing eligibility requirements for this status;
    c. liability exposures would be limited by provisions that clarified that the term 'owner or operator' did not include persons who hold title to a site solely as a trustee, custodian, or fiduciary, provided that they do not contribute to the release or threatened release of hazardous substances;
    d. settlement opportunities would be provided to municipal solid waste generators and transporters, with their aggregate liability limited to not more than 10 per cent of the total cost of cleanup;
    e. protection from future liability arising from new technologies or discoveries of environmental hazards would be made available, with EPA empowered to enter into complete and final covenants not to sue for future liability, provided the settlors paid a premium to the Fund for the risks of remedy failure and unknown conditions;
    f. the federal government would be exempted from liability when its ownership interest in a mining site resulted solely from its statutory land-manager functions, when it responded to disasters or threatened toxic releases, or in the event of economic regulation of industry during wartime.

6. Implementation of an obligatory early cost allocation process for NPL sites to reduce transaction costs and increase equity in the

allocation of cost shares. EPA would provide the opportunity for the allocation parties to voluntarily settle their cost shares and would accept fair and reasonable settlement offers based on the allocation scheme. Orphan shares would be covered by the Fund. PRPs willing to waive their right to seek contributions from other PRPs would be entitled to final release from future liability for remedy failure and undiscovered risks as an incentive to settle.

7. Acceleration of cleanups through promulgation of a national risk assessment protocol and national generic cleanup levels for specific chemicals to reduce inefficient site-by-site study. Target cleanup levels would reflect different land uses (e.g., residential versus industrial land uses). Sites located in states with more stringent cleanup levels would be remediated to the state levels.

8. Elimination of the current mandate for permanent remedies for those sites for which a treatment remedy is not available or is too costly when it is likely that a less costly treatment remedy would become available within a reasonable period of time, permitting the use of interim containment measures.

9. Provision of greater resources and time for Fund-financed removals, raising cost limits from $2 million to $6 million and the time for completion from one to three years.

10. Creation of an Environmental Insurance Resolution Fund, financed by fees on insurance companies, to provide insurance policy-holders comprehensive resolutions of their CERCLA claims, to reduce litigation between insurers and insured, and to expedite the availability of funds for response actions.

EXAMPLES OF CERCLA SUCCESSES

Throughout the US, there are thousands of examples of contaminated property redevelopments that have successfully generated jobs and revenues. Because there is no inventory of non-NPL projects, however, we must rely on selected case histories to draw lessons (e.g., Page and Rabinowitz 1994). The most comprehensive collection was assembled by Bartsch and his colleagues (Bartsch et al. 1991; Bartsch and Collaton 1994) who have gathered 18 case histories from 12 states. In

New Haven, Connecticut, for example, an old Winchester Arms Company was converted to a light industrial centre. In Commerce, California, an old Uniroyal Tire factory has been renovated to become a shopping mall, offices and a hotel. In Meadville, Pennsylvania, a former synthetic fibre-making plant has been converted into an industrial park.

The common themes running through the cases are, first, cooperative public–private partnerships involving developers, lenders, community organizations, state and federal regulators, and other local and state government officials. Second, the projects entail creative combinations of funding methods such as tax increment financing, state loan funds, joint ventures, and application of property cleanup tax credit (Boman 1991; Martin 1991a, 1991b; Swartz 1994). In some instances, cleanup efforts occurred as part of a large-scale urban redevelopment campaign in which a momentum was already in play when the pollution (or extent of the pollution) was discovered and had to be dealt with because of CERCLA-imposed liability. The case of Wichita, Kansas, provides an example (Tripp 1991; Glaser 1994).

As in many US cities, officials of Wichita were concerned about the degeneration of the central business district and had laid plans for revitalization. In 1990, however, the project was brought to a halt with the discovery of extensive groundwater contamination covering a six-square mile area beneath the business district. The contamination was travelling at about a foot per day and would eventually pollute wells in surrounding farmlands. More than 500 parties were designated as PRPs and cleanup was estimated at $20 million. In lieu of local remediation action, Superfund NPL listing was imminent. EPA pursuit of the PRPs and additional oversight costs threatened to add substantially to cleanup costs and to delay implementation of downtown regeneration. In response, the city manager and city council took charge, forming a coalition of local government and businesses.

Under threat of sweeping CERCLA liability, arrangements were forged to prevent the collapse of the revitalization plans. The primary contributor to the contamination, fearing suits over property value losses from the other PRPs, signed an agreement to assume a substantial portion of the damage costs. To secure backing from financial institutions, the city issued certificates of release from environmental liability to them, agreeing to accept responsibility for cleanup of the most seriously contaminated zone. With collateral thus assured, lenders agreed to finance the restoration. Other PRPs then cooperated to develop a formula for assigning their fair-share of cleanup expenses to avoid lawsuits that could have taken years. Orphan shares of remediation costs for polluters no longer in existence were funded by tax increment financing (that is, by the city borrowing against increases in its future property tax revenues expected from regeneration).

These cases underscore the importance of the individualist ethos in the US response to contamination. In the absence of the threat of EPA action, made greater by the very costs and delays criticized by those seeking CERCLA amendments, local initiative to mitigate contamination may be undermined by the inter-local competition for businesses and residents that prevails. In the absence of any sense of collective responsibility, individual property owners or businesses appear to have no incentive to cleanup either, and the CERCLA-induced threat of liability is essential to stimulating them to act. Without what has been called by some 'the 800 pound gorilla', of the Superfund, it is probable that in many instances the untended pollution would continue to contaminate and spread, threatening the ecosystem and human health.

PART III

5. The European Context and European Union Environmental Policy

We now shift our attention in this chapter and the next to Europe and more specifically to the European Union (EU). As a central jurisdiction, the EU parallels the US; as a supra-national body, it may forecast the future of NAFTA. In both cases, the EU provides a comparator to the North American, especially the US, experience.

The EU environment policy context and its underlying principles are important not only for those operating within EU member-states: the wider European scale encompasses countries with widely differing experiences of environment policy regimes, for most of which EU environment policy is actually or potentially applicable through enlargement following accession of new member-states, association agreements or technical assistance programmes. Given the enormous scope of this subject, all we can do here is to provide signposts and guidance through its complexities sufficient to explain the forces acting on policy formation and implementation.

In this chapter, then, we first review the unique nature of the EU as a supra-national jurisdiction, and its relationships with other parts of Europe and with its member-states. We then go on to examine the emergence of the EU powers to set environmental standards for member-states and businesses operating within the Union. Third, we turn to an outline of the processes of implementation of the EU environmental programme in general. We conclude with an overview of the rationale for a specific EU policy regulating contaminated land itself. (Chapter 6 provides a more detailed look at the extent and distribution of the contaminated land problem in Europe and examines in greater depth the EU law and policy towards contaminated land that exists or is in the process of evolving, as well as member-state practices.)

We thus trace the historical genesis of the current EU approaches to land contamination, covering treaties, changes in the composition of the Union, and the promulgation of key EU legislation. Table 5.1 provides a summary chronology of the developments as a reference for the discussion that follows.

WHICH 'EUROPE'?

The definition of 'Europe' has continued to change over the post-World War II period. What is now the fifteen nation European Union was not the first supra-national coalition; the alliances that confronted each other across the 'Iron Curtain' predated it. Yet the 1957 formation of the six nation European Economic Community (EEC) has proved to be the most significant political institution in shaping post-Cold War Europe. We look here first at the emergence of the EU since 1957, and then set that institution within the broader context of the other 'Europes'.

The European Union

The European Economic Community (EEC) was founded by the six original member-states who signed the Treaty of Rome in 1957, agreeing to create the EEC as of 1960. The organization has grown over time both geographically by successive enlargements, and in the extent of its powers and policy-making significance. It is qualitatively different from any other supra-national association of member-states, such as the Council of Europe, the Organization for Economic Cooperation and Development or NAFTA, all of which must rely on their members' governments to adopt and implement any policies they may have agreed to pursue. The unique feature of the EU is the extent to which it is a supra-national government and jurisdiction: European law takes precedence over national law enacted by the national Parliaments of member-states. Despite its growth from six to fifteen members, it does not equate with Europe as a whole. As we shall see, however, for certain purposes its policy regime extends beyond its boundaries to include other European countries.

Environmental issues were not even mentioned in the Treaty of Rome. Successive treaty revisions since 1957 have greatly added to the scope of the original Treaty, perhaps most notably with respect to the scope of supra-national environmental regulatory powers within which

Table 5.1 Selected events in the evolution of EU environmental policy

YEAR	EVENT OR POLICY ACTION
1960	Treaty of Rome (6 member-states, Belgium, France, Federal Republic of Germany, Italy, Luxembourg, The Netherlands) Foundation of European Free Trade Area (7 members)
1973	1st Enlargement: UK, Ireland, Denmark from EFTA (9 member-states) Agreement to establish an EEC Environment Policy
1976	Seveso disaster occurs in Italy
1981	2nd Enlargement: Greece (10 member-states)
1982	*Directive on Major Accident Hazards* (the Seveso Directive, EEC/82/501)
1985	*Environmental Assessment Directive* (EEC/85/337)
1986	3rd Enlargement: Spain, Portugal (12 member-states)
1987	Passage of the Single European Act, amending Treaty to formally incorporate Environment Policy
1989	Commission proposal for a *Directive on civil liability for damage caused by waste* (COM/98/282) Fall of Berlin Wall, revolution in Central and Eastern Europe
1990	Issuance of Commission *Green Paper on the Urban Environment* (COM/90/218) Agreement to create European Environment Agency (Regulation EEC/90/1210) German Reunification (3 October)
1991	Amended proposal for a *Directive on civil liability for damage caused by waste* (COM/91/219) Ratification of the Treaty of European Union (Maastricht Treaty)
1992	Single European Market in effect (as of 31 December) Adoption of *5th Action Programme on the Environment: 'Towards Sustainability'*
1993	Issuance of Commission *Green Paper on Remedying Environmental Damage* (COM/93/47) Maastricht Treaty in force (as of 1 November)
1994	European Environment Agency set up in Copenhagen
1995	4th Enlargement: Austria, Finland, Sweden (15 member-states)

any measures affecting contaminated land would be found. It was introduced into the Treaty by means of the Single European Act of 1987, the main purpose of which was to define the steps to be taken for the completion of the Single European Market (SEM, a common market of all member-states) as of the end of 1992. This Act provided the organization with explicit competence (i.e., the legal authority) over the environment for the first time by adding an Environment Title to the original Treaty and setting out powers and duties over the environment (Williams 1989).

Environmental politics moved onto centre stage in the period following the 1989 elections to the European Parliament, as the Environment Commissioner then responsible, Carlo Ripa di Meana of Italy, ensured that it had a high profile in the Commission's policy-making. It was in this climate that the negotiations resulting in the Maastricht Treaty forming the European Union and increasing the political and legal power of the supra-national body, took place in 1991.

The 1991 Maastricht Treaty (more properly the Treaty of European Union: since the Maastricht Treaty came into effect, the term European Union is now correct for what was previously the European Economic Community) thus enhanced the organization's environmental policy regulatory powers. This Treaty, which went into effect on 1 November 1993, is quite clear about the place that environmental considerations must have in all EU policy instruments and EU legislation. Title XVI, the Environment Title, sets general objectives to which EU policy on the environment must contribute. Most importantly its second paragraph states that, *inter alia*,

> Community policy on the environment shall aim at a high level of protection...shall be based on the precautionary principle and on the principles that preventive action should be taken, that environmental damage should as a priority be rectified at source and that the polluter should pay. Environmental protection requirements *must* be integrated into the definition and implementation of other Community policies. (Article 130r(2), EC Council 1992, authors' emphasis)

However, the Maastricht negotiations also introduced a new European key-word: subsidiarity. This is the doctrine that decisions should be taken at the most local level of government appropriate to the issue in question. Environmental policy was designated as one of the sectors to which subsidiarity is not only applicable but is to be publicly and demonstrably applied. Thus, in the absence of a clear demonstration that an environmental problem is 'European' in scope and requires specific EU action, policy initiative is left to the member-states (or, at their discretion, sub-national jurisdictions). Differences between member-states in their policy towards the environment, the interpretation they apply to EU environment policy obligations, and in their political will to pursue environmental protection objectives are becoming evident. Application of the subsidiarity principle, weakening the central commitment, combined with the departure from Brussels of the energetic environment commissioner, Ripa de Meana, took some of the impetus out of EU environmental policy-making over the 1992–94 period.

The pendulum may be expected to swing back again. Following the enlargement of the EU to 15 member-states in January 1995 with the accession of Austria, Finland and Sweden, the balance of power between the more and less environmentally concerned countries now shifts in favour of those seeking to adopt higher minimum standards. Given the treaty powers and the much wider applicability of 'qualified majority voting' than existed before the Treaty of European Union was adopted, the scope for any member-states wishing to block new environmental legislation will be much reduced. Another new development is the establishment in 1994, after some delays, of the European Environment Agency (EEA) in Copenhagen.

The EEA was authorized in May 1990 (Regulation EEC/90/1210, *Official Journal* 1990). That Regulation, however, came into effect only after a site for the Agency was designated, and agreement on the location in Copenhagen was much delayed, so the EEA only came into formal existence in 1994. The main role of the Agency, at least initially, is to assemble and monitor data, develop assessment methodologies, set up a network of national focal points, and provide the Commission with data and advice. Priority topics include air quality and emissions; water pollution; soil conditions; land use; waste management; noise; chemical hazards; and coastal protection. The Agency's impact on environmental policy as a whole — and on contaminated land and toxics policies in particular — has yet to become evident.

The New European Architecture

The EU is *not* 'Europe', operates within a broader continental framework, and relates to other trans-national organizations. To oversimplify what is sometimes called 'the new European architecture', four supra-national groupings of European countries can be identified, each with different environmental policy and problem profiles: the EU itself; the European Free Trade Area (EFTA), once seven countries and now reduced to four; central Europe (the Visegrad and PHARE groups); and the Commonwealth of Independent States (former Soviet Union) or TACIS group (Williams 1995b).

The Single European Market (SEM) came into operation on 1 January 1993, creating a common market in goods and services. A key element of the SEM was the extension of the so-called 'Four Freedoms', free movement of goods, services, labour and capital, throughout the member-states of the EC by the removal of all remaining tariff and non-tariff barriers. The member-states of the European Free Trade Area (EFTA) negotiated an agreement with the EC/EU to form what is the second main element in the new European architecture, the European Economic Area (EEA). The essence of this agreement is to extend the four freedoms to all EFTA member-states in order to create a much larger common market. The EEA is seen by the European Commission as a transitional arrangement leading to full integration in the EU, although this may not occur. As of 1995, the remaining EFTA countries are Iceland, Norway, Liechtenstein and Switzerland. The EEA agreement only partially covers EU environment legislation. Nevertheless, given the high environmental standards that all EFTA countries set themselves, the whole of the EEA can be regarded as having EU standards as a minimum.

Since the revolutionary events of 1989, a major EC/EU priority has been to establish political and economic links with the former communist party states of Central and Eastern Europe (CEE). All have had the experience of regimes which have paid no attention to concerns over pollution, air and water quality, or conservation of the natural environment. The Gölitz–Usti–Katowice triangle in the border regions of the former German Democratic Republic (GDR), Silesia in Poland and northern Bohemia in the Czech Republic contain some of the worst polluted environments and problems of contaminated land in Europe if not the world.

These countries are now attempting to overcome this legacy, supported in part by EU aid programmes which include recovery from pollution and land contamination among objectives eligible for funding. The central Europe group includes six countries aspiring to EU membership, Bulgaria, Czech Republic, Hungary, Poland, Romania and Slovakia, all of which are attempting to bring their environmental standards up to the minimum required EU levels in order to become eligible to join the Union.

The EC has already had some experience of accommodating the environmental legacy of communism and addressing it in policy development, following German reunification in October 1990. Although not officially an enlargement, since an existing member-state was adjusting its boundaries and there was no Treaty of Accession, in practical terms the EU acquired a new geography and a whole set of new problems, environmental as much as political and economic, as the Bitterfeld case discussed in Chapter 6 illustrates. It is thus extremely sensitive to the problems of the CEE nations and to the extent to which their environmental conditions can affect EU member-states and their citizens.

The new European architecture is therefore taking shape with three main building blocks:

1. countries now in the EU of 15 member-states or the remains of EFTA forming a western crescent of an enlarged EU from Greece via Western Europe to Finland;

2. a Central Europe of countries benefiting from PHARE with close political, economic and infrastructure links to EU member-states, and aspiring to EU membership;

3. the countries of the former Soviet Union in the CIS, benefiting from the TACIS Programme, but with a much longer road to follow in their transformation process.

Some of the boundaries between these groups may not be certain yet, and some countries may fall outside these broad blocks, but the macro-scale political geography of Europe can usefully be simplified in this way.

EUROPEAN UNION ENVIRONMENTAL CONTROL POWERS

Unlike most national governments, the EU cannot legislate unless it has the specific legal power or competence to do so. This power comes from the treaties, and all proposals for legislation must be explicitly tied to an article in a treaty. Notwithstanding the age of the original

European Community, its experience operating with the broad priorities and mandate for environmental policy formation contained in the Maastricht Treaty is minimal: the EU has operated under that treaty only since November 1993. It is the duty of the Commission to make proposals for legislation, as well as to monitor and administer existing policies. In developing an approach to environmental policy, the Commission will, presumably, be assisted by the new European Environment Agency, but that entity became operational only in 1994. Thus, the environmental policy powers the EU now wields have not really been tested or applied.

European law is not made by the Commission, however. The body which makes the law is the Council of Ministers, consisting of one member from each member-state. When the Council is sitting as the Environment Council, that is, with proposed environmental legislation to consider, it consists of the respective national Ministers for the Environment. Before the Council can act upon a proposal from the Commission, or enact a draft legal instrument, the opinions of various consultative bodies must be obtained, the most important of which is the directly-elected European Parliament. That body, in turn, is also in flux, with the 1995 enlargement of the EU from 12 to 15 member-states.

The three types of EU legal instruments, Regulations, Directives and Decisions, may all be employed in environmental policy-making; to date, environmental *Directives* have been most frequent. They are legally binding as to the ends to be achieved (the legal requirements which must be incorporated into national law are specified by the superior European law), but the manner in which this is to be accomplished is left to the member-states. Directives are normally adopted whenever an EU measure relates to a topic on which a body of law already exists in the member-states; the variability in the individual national laws is such that the details of amending them to conform to the EU legislation are simply not worth the attention of the supra-national body. A *Regulation* is legally binding in its entirety: the text as written enters national law directly; this instrument is normally used when an altogether new subject is legislated and it is not necessary to integrate the text with that of existing national law. A *Decision* is also legally binding but, unlike the other two instruments, is addressed to individual legal bodies or private persons, not member-states.

Proposals from the Commission in the form of draft Directives, Regulations or Decisions may be adopted under a rule of unanimity (which provides each member-state with a veto) or by a procedure known as Qualified Majority Voting (QMV). The Maastricht Treaty of European Union has made it possible for the great majority of environment measures to be submitted under treaty powers to which QMV applies. Under this system, each member-state has a number of votes, very roughly proportionate to size:

- France, Germany, Italy, United Kingdom — 10 votes;
- Spain — 8 votes;
- Belgium, The Netherlands, Greece, Portugal — 5 votes;
- Austria, Sweden — 4 votes;
- Denmark, Finland, Ireland — 3 votes;
- Luxembourg — 2 votes.

With 62 out of the total 87 votes required for a majority, a blocking minority coalition now requires 26 votes and must normally therefore have at least three large countries, or two large and two small countries in order to succeed. Previously, three countries of any size were often enough to form a blocking minority. The enlargement has thus not only added environmentally concerned member-states to the EU, but diluted the powers of those nations reluctant to see stricter regulations to block such actions on the part of the supra-national body.

The importance of this shift in power is clear from the content of the Environment Title of the European Union Treaty. Article 130r, quoted above, is concerned with general objectives and principles, and places the environment on centre stage as a 'constitutional' priority. Article 130s states that proposals on all environmental matters except for a specified list of topics may be adopted by qualified majority voting. This article is noteworthy for the inclusion of 'measures concerning town and country planning [and] land use' among those requiring unanimity. Contamination, with possible off-site consequences, may, arguably, not be a matter of land use, but one of waste or product disposal, which would be subject to QMV. In due course, this broad impact is likely to be used to enhance control over polluting land uses. Article 130t allows any member-state to adopt more stringent environmental standards than those laid down in an EU directive or regulation. This was included at the insistence of Denmark, and is also needed to satisfy Finland and Sweden (Budd and Jones 1991; Lodge (ed.) 1993; Nugent 1991).

Principles of EU Environment Policy

Certain basic principles have been laid down in successive statements of EU environmental policy, as we have noted: the polluter should pay, prevention is better than cure, the sustainability of practices is important, and action should take place at the appropriate level of government. Each underlying precedent has shaped policy.

That *the polluter should pay* is widely accepted in principle though sometimes difficult to implement in practice. The greatest challenge arises in the area of former communist countries where the polluter may no longer exist (e.g., state enterprises of the former German Democratic Republic) or have no possibility of being able to pay the sums necessary to overcome the environmental legacy of years of neglect. For this reason there is substantial EU interest in the US Superfund concept (UK DOE 1994).

The principle that prevention is better than cure, or the *precautionary principle* has been an aspiration since 1973, although many of the earlier measures responded to the need to give priority to remedial action against existing pollution. Adoption of this principle is made explicit in Article 130r(2) of the Treaty (O'Riordan and Cameron 1994). The logic of prevention and the precautionary principle led the Commission and its environment directorate to focus attention on shaping and influencing the land-use planning process, since all systems have a procedure for the authorization of development which takes place at an appropriately early stage of the decision-making process. The Environmental Assessment Directive of 1985, which is integrated with the land-use planning process, is an excellent example of the application of this logic.

Sustainability raises philosophical and conceptual issues, although the word is also used often, like subsidiarity, for reasons of political expediency. The accepted classic definition of environmental sustainability is that first given in the Brundtland Commission report: 'development that meets the needs of the present without compromising the ability of future generations to meet their own needs' (Brundtland Commission 1987:8). Since the publication of that report, and the environmental summit in Rio in 1992, the word has become common currency in environmental discourse and policy statements. Article 2 of the Maastricht Treaty, concerned with the basic principles of the EU, sets as one of its tasks maintenance of 'sustainable and non-inflationary growth respecting the environment'.

Subsidiarity has come into prominence since the Maastricht summit. It is a principle that is intended to apply to all policy sectors where implementation depends on the actions of national or sub-national institutions. In the environmental policy context, a sector in which there is a clear political intent that the principle should be applied, subsidiarity should be taken to mean the long-established EU principle that action should take place at the appropriate level of government, whether that is at the local, regional, national or supra-national level. Interpretation of this is a matter for individual member-states. It is seen by some as no more than an elementary rule of good government. However, the term has been adopted and distorted by those, such as the UK government, who argue for more power to be assigned to national rather than sub-national or supra-national authorities.

Why an EU Environment Policy?

Why should an economic common market (which is what the EEC originally set out to be in 1957) have a centrally-dictated environment policy? Initially, the need for such a policy-sector was not recognized, and an effective environmental policy came into being only in 1973. Several justifications have been put forward, based either on quality of life considerations or on the positive economic benefits deriving from environmental action (Williams 1990). Clearly, political pressures from the electoral strength of Green parties also played a part.

Given the international and cross-border nature of many environmental issues and their high political salience, the principle that the EU should have environmental regulation among its competences may seem obvious today, but in early statements and action programmes on the environment it was felt necessary to address this question. Essentially, the rationale for an environment policy in an economic body such as the EU is based on three sets of arguments: (a) the competition argument; (b) the quality of life argument; and, (c) the economic benefit argument.

The competition argument is the fundamental reason, connected with the creation of the single market, the adoption of the four freedoms, and the consequent need to remove non-tariff barriers and any form of distortion of competition alongside the more obvious programme of removal of all tariff barriers at internal borders within the EU. This is sometimes expressed by the metaphor of creating a level playing field (appropriate to either the American or World version of 'football'). Variable environmental and pollution control standards between different member-states are seen by the European Parliament and the Commission as a form of non-tariff barrier that could distort competition and consequently interfere with the proper functioning of the common market (Fairclough 1983).

Initially, as the first steps towards the Single European Market were taken, the danger of the emergence of 'pollution havens' creating such distortions was not fully appreciated. The necessity of thinking in terms of the EU as a single jurisdiction, one market and one economic space was not adequately appreciated by the representatives of the member-states. This myopia is well illustrated by the position taken by several organizations in Britain opposing the proposed *Directive on Environmental Assessment of Projects* on the grounds that the UK already had legislation allowing its objectives to be met (such as the Town and Country Planning Acts), and therefore had no need of European legislation (Williams 1983, 1986). This argument completely missed the point of the idea of a common market. The five-year long SEM implementation programme leading up to the 1993 launch of the single market and its four freedoms was instituted in large part to give time for these calcified perceptions to change.

The quality of life argument can be expressed in either social or economic terms. Those emphasizing the social viewpoint argue that the EU does not exist solely to benefit business and governments, that all EU citizens are entitled to enjoy the benefits of a high standard of living, and that high environmental standards thus are essential. The economic emphasis is based on the argument that quality of life criteria affect location decisions made by entrepreneurs, whether or not the industrial or commercial activity being undertaken is directly dependent on a clean environment, since the decision-makers would choose to locate, other things being equal, in cities or regions with an attractive environment offering good living conditions.

The economic benefit rationale goes beyond the level playing field issue to recognition of the sectoral implications of environmental standards. For certain sectors of the economy, notably agriculture, but including some manufacturing processes, attainment of satisfactory operating conditions is directly dependent on the maintenance of environmental quality standards. In others, the markets for certain manufactured products or services may be stimulated by demands for high environmental quality conditions. The EU recognizes that, in both instances, overall *economic* well-being may be enhanced by demands generated either by government regulation or societal expectations for more stringent *environmental* standards.

Responsibility for EU Environmental Policy-Making

Formal responsibility for preparing policy and proposals for legislation rests with the European Commission. The division of the Commission responsible for environment policy is the Directorate-General for Environment, Consumer Protection and Nuclear Safety, DG XI in the normal shorthand, under the political direction of a member of the Commission. Several other DGs also develop policies with significance for the environment, among them agriculture, energy, transport and regional policy.

Draft proposals are subject to extensive consultation both in public and in private. One environment Directive, the 1985 Directive on the environmental assessment of development projects (Directive EEC/85/337, *Official Journal* 1985) was reputed to have been the subject of more consultation and revision than any other Directive on any subject, partly in order to meet French and British objections. When it was believed that this was accomplished, and the way clear for adoption by the Council of Ministers in 1983, it was vetoed by Denmark on the grounds that it was too weak and ineffectual. It took until 1985 for the Danish veto to be lifted (Williams 1986). The move towards the SEM stimulated the growth and influence of the increasingly extensive network of lobbying organizations in Brussels representing industrial and environmentalist interests is not to be discounted. They normally are formally consulted, at least by key committees of the European Parliament, so their influence continues to grow.

IMPLEMENTATION OF THE EU ENVIRONMENTAL PROGRAMME

Notwithstanding the short lifetime of the EU as a supra-national body, its predecessor organizations, the European Economic Community and the European Community, as well as its member-states, have implemented environmental legislation and regulations. Policies and an explicit series of programmes for change have been promulgated, with varying levels of impact.

Modes of EU Policy Implementation

The Commission has, since 1973, adopted the practice of setting out a multi-annual environment action agenda in the form of successive *Action Programmes on the Environment*; the EU is now operating under the fifth such instrument. The status assessments, reviews and statements of policy development intended for the period covered by each Action Programme, are formally non-binding. However, many proposals have been adopted as EC or EU law, as either Regulations or Directives. Haigh (1989) listed over 200 items of European environmental legislation, and the rate of adoption has increased from around 20 per annum in the 1980s to over 30 annually in the 1990s (Williams 1990). Many of these laws were minor or narrow in scope, but the accelerated pace of EU environmental legislation speaks to the growing acceptance and recognized importance of the issue.

The current *Fifth Action Programme*, for the period 1992–2000, is entitled 'Towards Sustainability' (CEC 1992a). This is the most comprehensive to date and is published in three volumes. It highlights as target sectors industry, energy, transport, agriculture and tourism, and develops the theme that industry should be seen as part of the solution rather than only as a source of problems, emphasizing subsidiarity and the need for bottom-up policy development.

The Commission also adopts the practice of issuing 'Green Papers' to indicate policy thinking on which consultation and reaction is sought prior to promulgation of specific law. Green Papers set out longer-term visions which do not always meet with sympathy or understanding by the more pragmatic Anglo-Saxon mentality (as exemplified by the British media including the technical and professional press). A far-reaching example is the *Green Paper on the Urban Environment* (CEC 1990), which sets out proposed urban planning principles that would meet environmental and energy-efficient objectives. It did not explicitly address the contaminated land cleanup question, although its planning and policy principles were designed to have the effect of avoiding such problems in the future.

The difference between a rhetorical or visionary and a pragmatic style of policy-making is worth noting with respect to the use of such instruments, since it is reflected in member-state responses to the Green Papers. Over-generalizing, we may associate the former with French or possibly Italian thinking, the latter with the Anglo-Saxons. The different outcomes are most starkly illustrated by the saga of the Channel Tunnel and associated infrastructure: the French have moved quickly from spatial conceptualization to a national strategy for road and high-speed rail links which were in place as the 'Chunnel' opened; the British were still struggling to appreciate the nature of the project and the scale of the response necessary, beyond construction of new rail termini (cf. Halliday et al. 1991; Williams 1991). It is not really possible to attribute either of these 'pure' models to the other EU member-states; most exhibit characteristics of each mode of policy thinking.

Trans-National Urban Competition and Networking

The advent of the SEM has led to a growing sense of spatial competition between cities and regions throughout Europe as a whole: to quote a former Italian Foreign Minister, 'the single market is not so much the competition between companies as the competition between systems, and those countries and regions with the most appropriate systems will be the winners' (de Michaelis 1990). Associated with this mode of interaction are the phenomena of city imaging and marketing, and of network-building by local and regional authorities.

The RECITE (Regions and Cities of Europe) programme of the Commission gave financial support to over 30 networking projects, some of which were *ad hoc* groups of city or regional authorities and others were, or became, permanently established associations. None of these are directly focused on contaminated land development as such but for several, sharing experience and advice on reclamation, and promotion of development on formerly contaminated sites, figure among the objectives. Eurocities, an association of major regional cities which is proving to be quite influential in urban affairs in the EU, is a case in point. The Association des Régions Européennes de Technologie Industrielle (RETI) is particularly important in relation to contaminated land, since it is an association representing regions with a history or legacy of heavy coal- and steel-based industry; it conducted a Europe-wide survey of contaminated land, to which we return in Chapter 6.

Planning and Land Policies in the Member-States

All European countries have their own systems of land use and spatial planning, their own traditions and practices of land policy, tenure and property ownership. All have developed in the specific national legal and cultural context, in some cases long before the question of

European integration arose. Now, these systems can be said to represent non-tariff barriers and therefore impediments to the free operation of the SEM. The response of the professionals and development industry has been to develop the expertise to handle this diversity rather than to press for closer harmonization of procedures. Two dimensions of difference stand out as almost insurmountable, defying integration.

First is the distinction between the legally-binding and the discretionary systems. Broadly, the system in Roman law countries is based on the principle of legally-binding plans that are not subject to modification without amendment of legislation. Two member-states have discretionary planning systems, along with common law legal systems. These are the UK and Republic of Ireland, where the planning systems allow consideration of each development proposal on its merits within the context of a planning policy set out within a broad framework rather than the detail that is characteristic of the legislation in legally-binding systems. There is less certainty and more flexibility as it is possible for the planning authorities to depart from the development plan if justified by the merits of individual proposals.

Second, is the extent of government control of land markets, especially with respect to land policy and tenure. The spectrum ranges from the relatively free land markets of the UK to the highly managed system of The Netherlands. In the UK, it is accepted that property development may be speculative and may be undertaken by those whose interest is investment rather than as users of the development. In locations where there is not a strong market, a typical situation in regions where contaminated land is most widespread, intervention may take place in order to create market conditions in which the private sector could find it attractive to invest. Ironically, such intervention has resulted in an expansion of plan-led development activity over the course of the Conservative governments in the UK since 1979.

By contrast, public authorities in The Netherlands are expected to make land available for development when and where required. The public obligation is akin to the duty of a water supply utility to supply water wherever and whenever it is requested (Needham and Kruijt 1992). Municipalities supply land on sites consistent with their development plans, on which industrialists develop their premises as they require. Land values are very stable and predictable, varying little from place to place. Investors have little to gain from development activity on industrial property. The property developer as speculative risks taker is absent from Dutch land markets, since all that is required for acquisition of developable property is a visit to the town hall.

Other countries fall between these two extremes. In Germany, for example, which has relatively low levels of residential owner-occupation, it is rare for industrial occupiers to be anything other than owner-occupiers. An exception is the Hansapark in Düsseldorf, a private development of prime rented property in one of Europe's most prosperous cities. The extent of public–private partnerships for development or reclamation further modifies these polar examples.

THE RATIONALE FOR AN EU CONTAMINATED LAND POLICY

On the basis of the subsidiarity principle, there is clearly a case for arguing that land pollution, unlike water and atmospheric pollution, is a matter for national rather than EU action. However, one overriding issue that has clearly emerged in the US (Yount and Meyer 1994a, 1994b) and increasingly in Europe now (*The Economist* 1994; Moore 1994) is that of the assignment of liability for environmental damage. Contaminated land, and fear of financial liability associated with it (whether for cleanups or for environmental or human health damage), has become an issue of major concern to developers, financial institutions investing in development and policy-makers concerned with regeneration of urban brownfield sites.

Undue fear of the risk of liability for any contamination may be deterring developers from sites where there is any chance contamination may be found to exist. To the extent that concern over contaminated land is an issue capable of distorting the operations of the land and development or investment and insurance markets, it is a single market issue and therefore clearly an appropriate subject for EU action. This argument that subsidiarity may not be applicable to contaminated land policy has even been acknowledged by the UK, the strongest advocate of the subsidiarity principle (UK DOE 1994, para 4B.20).

A second impetus for an integrated contaminated land policy derives from the high level of EU concern with questions of overall quality of life and of the urban environment. The logic of the *Green Paper on the Urban Environment* (CEC 1990) and of national policy-making in several member-states is to encourage reuse of brownfield, or formerly urbanized, development sites in order to avoid urban sprawl and excessive conversion of greenfield land. Thus EU action to *support* national planning objectives and standards may be warranted in light of the competitive pressures of the SEM.

A third factor is the issue of the possible future enlargement of the Union to include Central European countries. All the candidate countries in this group have a legacy of contaminated land as a result of polluting activities in the past which are of a vastly greater order of magnitude than those typical of even the worst examples in the present member-states. Thus a central policy and standard, and provision for financing cleanups, may be required to address the current and future inequalities generated by these past acts of contamination.

The EU is still in its infancy and is again in flux with a recent enlargement. Prediction is hazardous under such conditions. Yet the logic underlying the past environmental regulation and law in the EEC and EC, combined with these three elements of potential distortion of the SEM associated with the market consequences of past pollution suggest that it is only a matter of time before an integrated EU contaminated land policy is promulgated.

6. Contaminated Land Policy in the European Union and its Member-States

We turn now to an examination of contaminated land policy within the context of broader EU environment policy, discussing indirect and direct measures and proposals from the European Commission and process of policy delivery. We explore the diversity that is a dominating feature of the EU, colouring the different perceptions of the extent to which contaminated land is seen to be a problem.

Inevitably, no more than an overview of the whole of the policy picture in Europe is possible. Our aim is to examine the extent to which there is a European view as distinct from a series of national pictures, and to explore the extent to which contaminated land is addressed in EU policy-making, in order to provide the basis for our comparative analysis. We begin with a description of the nature of the overall European contaminated land problem, in order to provide a background picture. Next we present an overview of the issues in contaminated land policy that have been prominent in the EU debate, addressing commonalities and differences between member-state approaches as well as the EU-wide debates. Third, we turn to the specifics of the EU policy development efforts to date, examining both Directives and Regulations promulgated and documents reflecting intents and future directions. Then we turn to the experience and policy developments in selected individual member-states in order to illustrate differing national positions and also the diversity of development processes associated with contaminated land. As we examine policies in different countries, we offer some examples of regeneration efforts to date. We conclude with a synthesis and some conclusions on the directions that the EU appears to be taking.

THE CONTAMINATED LAND PROBLEM

We have previously argued that past pollution of property is both a physical and biological (or 'technical') problem and a social and perceptual (or 'political') problem. The extent to which contamination is a major issue, then, varies with the extent of both problems in a nation-state. As we examine, first the 'technical' issue, and then the 'political' one, we note that the imperatives for action and national policies result from the combination of the two problem types.

The Technical Problem: Past Pollution

The countries of Europe can, by and large, be classified into two groups on the basis of their levels of known past activities and the extent of current land pollution. Our classification is based on one of the few studies to have attempted to estimate the scale of the problem on a comparable basis for the whole of the EU. This assessment was conducted by a non-government association of former heavy industry regions in Europe, the Association des Régions Européennes de Technologie Industrielle (RETI, European Regions of Industrial Technology) for the European Coal and Steel Community (RETI 1992). The report thus concentrated on contamination of former coal and steel sites, so the actual figures quoted must be considered to be substantial underestimates of the overall totals. They do provide an undoubtedly conservative picture of non-military contamination.

Evidence of the diversity of the problem across member-states is provided by the data on 'open-cast' mines, a modest issue in most member-states such as the UK where good restoration practice is normally required by the planning authorities. However, Germany's extent of problematic mine pits was found to amount to 70,000 hectares, substantially more than the total of 51,960 hectares of contaminated land identified for the whole EU in other problem categories. This problem is an environmental legacy of the communist regime in the former GDR, in which toxic wastes from chemical and other industrial plants were indiscriminately dumped in massive abandoned open-cast lignite mines.

Taking the RETI data on steel- and coal-related contamination to be indicative of the total volume of land pollution, the EU member-states and aspiring members fall into:

- *Group 1*: Five EU member-states, Belgium, France, Germany, Spain and the UK. They are the countries in which the bulk of the problem is found and include the countries with the earliest experience of heavy industrialization. They encompassed 94 per cent of coal- and steel-based contamination in the EU of 12 members (RETI 1992).

    The Central European former communist party countries aspiring to EU membership: Bulgaria, the Czech Republic, Hungary, Poland, Romania and Slovakia. Although not equally polluted, these nations have within their borders some of the worst examples of contaminated industrial lands to be found anywhere.

- *Group 2*: The remaining seven current EU member-states addressed by RETI, Denmark, Greece, Ireland, Italy, Luxembourg, The Netherlands and Portugal. We can add to their number the three countries that joined the EU January 1995, Austria, Finland and Sweden, who also have only a limited land contamination problem.

Among the countries of Group 1 above, the RETI study found over 12,000 hectares in UK, over 8,000 each in France and Spain, and over 6,500 hectares each in Belgium and Germany (excluding the open pit mines) contaminated. No Group 2 EU country had over 1,000 hectares contaminated: The Netherlands, Italy and Luxembourg were in the 650–950 range, with Portugal much lower, and Denmark exhibiting negligible land pollution. No data were provided for Greece and Ireland, but their past economic activities do not suggest a likelihood of extensive contamination. The overall estimate of around 52,000 hectares of contaminated land for the EU, excluding the open-cast mines, is less than many had feared. However, it *is* an understatement, and is further questionable due to national differences in variable definitions and perceptions of what constitutes contamination.

Internal inconsistency in findings are also a problem, as Table 1.1 demonstrated. A 1988 survey in Wales, an area of concentrated coal mining and steel production, found 746 potentially contaminated sites with an estimated area of 4,080 hectares, over a third of the RETI estimate for the UK as a whole, but domestic estimates for the entire UK range as high as 100,000 sites and 200,000 hectares (UK DOE 1994: par.2.10; Warren 1994). Studies of the Ruhr valley in Germany have been wildly inconsistent: a 1985 survey found 4,422 hectares of derelict land across all the region's municipalities, while another in 1988, covering less than one-third of the territory, reported a 50 per cent greater 6,570 hectares (Pohl and Grüssen 1991:281).

The Political Problem: Perceptions of Contamination

The legacy of contaminated land has not always been fully acknowledged as a consequence of industrialization and, to the extent that this was recognized, sometimes has been regarded as a necessary and inevitable concomitant of an industrial economy. Recognition and acceptance of the extent to which a debt is due for past pollution varies substantially across the nations of Europe. In some measure, variation across some of the dimensions introduced in Chapter 2 may account for national differences.

The extent of contaminated land and the availability of unpolluted sites for development are significant factors. The debates at the EU level as well as national policies suggest that the smaller member-states of the EU are more concerned about contamination mitigation than are their larger neighbours.

The strength of the public policy imperative placed on reusing brownfield sites affects the priority placed on mitigation as well as the economic returns to redevelopment investments. Annual loss of farmland to urban development, one measure of greenfield loss, varies from insignificant to over fifteen square metres per person per year across the EU member-states (CEC 1992b).

Widely publicized environmental 'disasters' have contributed to environmental concerns. Events such as the 1976 Seveso leak of trichlorophenol that contaminated 1,800 hectares and caused adverse health consequences in Italy, and the 1986 chemical plant fire in Basel that contaminated the Rhine river basin, thus created particular regional sensitivities to contamination problems, but also, through the EU, stimulated national responses to threats of contamination.

The differing financial systems and insurance availability in the member-states shape the political problem posed by past pollution. Countries, such as the UK, that lack publicly-oriented or nationalized financial systems are more inclined to define the contaminated land issue as one of financial risk and liability than other nations less fearful of capital abandonment.

The countries with the strongest environmental movements are those of northern continental Europe: Denmark, Germany, The Netherlands and Sweden. Of these nation-states, only Germany has a large-scale legacy of heavy industry and contamination. While the German case is exceptional in its inheritance of the legacy of pollution as a result of reunification, the strength of its Green Party predates its 1990 acquisition of the former communist state's contamination problems.

By contrast, the UK is giving closer attention to its contamination problem, but appears to emulate the US and differs from the rest of the EU in defining the problem primarily in terms of investors' financial risk and liability. Public policy efforts to mitigate pollution have been embedded in broader subsidy programmes for derelict brownfield lands. The national government disregard for health dangers, however, may be difficult to sustain, given the wide extent of unrecorded contamination and likelihood of problems. Although the Green Party has not been a factor in domestic elections, major environmental groups in the UK recorded membership gains averaging over 100 per cent over the decade of the 1980s, with activist groups such as Greenpeace and Friends of the Earth growing from a combined membership of 22,000 to 420,000, a pattern that reflects growing concerns (Gandy 1993). Further stimulus to growing concerns arises regularly. A 1994 article headed 'New tests at "Poison" Nursery', for example, detailed unexplained staff and neighbouring residents' ailments, no prior site use for manufacturing, but location in a 180-year old industrial area, suggesting unrecorded indiscriminate dumping (Eaton 1994).

As we suggested in Chapter 1, France has yet to develop a systematic approach to land contamination. The Netherlands so controls its domestic land markets that a central state role in mitigation is not a matter of political dispute: it is presumed, given the state obligation to provide land for development. Spain, at least around Catalonia, appears to be ignoring all too many of its problems, largely because of its inability to cope with NIMBY problems (Marshall 1993).

Further indication of the strength of environmentalist concerns may be derived from the extent of member-state 'compliance' with Commission Directives. The term compliance refers in this case to the passage or modification of national laws to comply with EU Directives, and is the only indication of willingness to adopt environmental regulations now available. (Practical compliance, that is, enforcement of environmental legislation, is beyond the Commission's capacity to monitor or enforce; while the European Environmental Agency was intended to perform this role, member-state governments confined themselves to monitoring environmental conditions rather than enforcing EU environment policy.)

There remain considerable differences between member-states in formal compliance implementation of EU Directives. Denmark is the only member-state where compliance rate for environment Directives is better than the average rate for all Directives. The other countries with good rates are Germany, The Netherlands, Spain and Portugal (although the latter two were relieved of some obligations, receiving so-called 'derogations' as part of their membership entry negotiations). Italy has the worst record, followed by Greece; the UK is in a middle group with Belgium, France, Ireland and Luxembourg (Wurzel 1993).

The Public Policy Imperative

The 1992 EU assessment of *The State of the Environment in the European Community* provides an excellent example of the intersection of the technical and political problem definitions (CEC 1992b). The data reported in the document were gathered to provide a rationale for the EU's Fifth Environmental Action Programme and represented the best that the Environmental Directorate could provide. Reliance on self-reported data, however, produced anomalous reports on the numbers of contaminated sites. No data on total sites were available for Belgium, Greece, Ireland or Luxembourg, despite the known high volume of past pollution in the former. Italy reported over 5,400 contaminated sites and Portugal over 1,800, which seems reasonable. The Netherlands, however, admitted to over 6,000 polluted properties and Denmark reported over 3,000, but France acknowledged only 800, the UK only 300, and Spain a mere 94 instances of contaminated lands! Germany, on the other hand, provided a figure of 35,000 contaminated sites to the EU data collectors.

These differences are not plausibly attributable only to variations in definitions: they reflect problem recognition and reporting tendencies. Combining the findings on technical and political problem definitions, we see that the latter tend to override the former. In the absence of environmental movements or Green parties that wield domestic electoral power, it appears as though the problem of land contamination, however extensive it may be in physical terms, has, at least prior to the advent of the Single European Market, been relatively ignored as a public policy matter.

A recent comparative analysis concluded that 'the German political system seems to be delivering far more effectively in terms of environmental policy-making...than does the British system' notwithstanding the fact that 'there are no convincing signs that the

German electorate is more ready than the British to change expectations and lifestyles' (Gordon 1994:11). This difference may be most readily attributed to the sustained presence of the Green Party as a political force in Germany.

The Greens have been represented in the European Parliament since 1984, but the EU focus has been on the forms of pollution with a more obviously cross-border dimension such as air and water pollution. The idea that contaminated land is an issue with a European dimension has not yet been accepted by all member governments (Williams 1995b). The debate over an EU policy has taken less environmental than economic dimensions, as the issue has come onto the agenda with the advent of the SEM. The argument for explicit EU policy rests on concerns with distortion of the single market by making investment in development projects on formerly contaminated land more attractive, or subject to less risk, in one member-state as opposed to another. Most member-states offer incentives to develop on brownfield sites as opposed to greenfield sites in the same region that could further distort competition. (It is further noteworthy that such subsidies are far more prevalent in the EU than in the US.)

To the extent that an explicitly environmental EU policy towards contaminated land is evolving, it appears to be based on 'technological' considerations. The latest enlargement, adding Austria, Finland and Sweden to the EU, may shift the focus, however. These nations have set themselves the highest environmental standards domestically and can be expected to exert their influence in the Council of Ministers in favour of setting high European standards. As a result, they may stimulate formal adoption of a number of EU policy initiatives that have not so far come to fruition (*Official Journal* 1991a, 1991b; CEC 1993a).

CONTAMINATED LAND POLICY DEVELOPMENT IN THE EU

Contaminated land is not identified explicitly as a sector of EU environment policy or in the terms of reference of the new European Environment Agency (authorized in 1990) and no Directive has so far been adopted explicitly addressing the problems associated with it. Yet the issue debates we have just reviewed suggest a preponderance of opinion and existing practices attesting to the logic of EU action. We examine here the history of EU efforts to frame an explicit contaminated land policy and the forces arrayed against its implementation.

The Commission first put forward a proposal in 1989 for legislation addressing contaminated land, the *Proposal for a Council Directive on Civil Liability for Damage Caused by Waste* (CEC 1989). The text appears to be derived in large part from CERCLA. The US law was then almost a decade old, and the EU has tended to look at earlier legislative experience in the US when appropriate. After debate and consultation with member-states, an amended proposal was put forward in 1991 (CEC 1991; *Official Journal* 1991a). The 1991 proposed Directive remains on the docket of the Council of Ministers, facing opposition from, among others, both Germany and the UK.

The amended 1991 version retreats somewhat from the 1989 text with respect to third party (including lender) liability, and introduces the concept of 'impairment to environment', a term which translates better into other EU official languages. The proposed text still seeks to establish strict joint and several liability rather than fault-based liability as a basis for liability claims, although with limitations. Under Articles 6 and 7 strict joint and several liability would be limited in instances of proven fault on the part of one or more parties. Article 10 limited liability to thirty years, protecting parties from claims for century-old damage. Finally, from the outset, Article 11 proposed the conduct of a feasibility study on the establishment of a 'European fund for compensation for damage and impairment to environment caused by waste'. This last element represents a significant departure from the US CERCLA approach, which accumulates funds for mitigation, but not to pay liability claims.

The year 1991 also brought a new proposal for a 'Directive on Landfill Waste' (*Official Journal* 1991b). The aim of this legislation was to establish guiding principles for location, design and operation of landfills, define what waste products could be deposited in them, and to provide a framework for funding of aftercare and maintenance following closure of a landfill site, in the form of a proposal for a Landfill Aftercare Fund (Article 18). Major conflict arose over the proposed phase-out of the practice of co-disposal, by which hazardous wastes are landfilled along with municipal solid waste and other non-hazardous material. The proposed Directive has been stalled by the UK, Greece, Ireland and Portugal, who objected to this provision (Eduljee 1994). Since this Directive is subject to qualified majority voting (QMV), indefinite blocking is unlikely to be possible since the four objecting countries no longer have the votes to block the majority in the fifteen-nation EU.

In 1993 the Commission issued its *Green Paper on Remedying Environmental Damage* (CEC 1993a) and the Council of Europe drew up its *Convention on Civil Liability for damage resulting from activities dangerous to the environment* (CE 1993). Clearly this indicates recognition that it is a widespread European problem. The Commission Green Paper takes as its basic premise the argument that different systems of civil liability could lead to distortions of competition, although in the view of sceptics this is asserted rather than demonstrated to be valid. It does not make precise policy proposals but is instead a wide-ranging discussion of the issues. It may be interpreted to be a retreat from the two 1991 Council Directives referred to above. However, it clearly has stimulated new debate on the policy questions and the issue of subsidarity.

Given the new QMV formula under the 1995 enlargement, the balance of power may have shifted towards the member-states that favour EU action on land contamination, so the additional debate and publicity associated with the Green Paper may, on balance, have been a step forward. We turn now to the elements and roots of the major EU policy debates on approaches to contaminated land.

ISSUES IN THE ONGOING EU CONTAMINATED LAND POLICY DEBATE

Europe contains locations such as north-east England, the Ruhrgebiet and parts of southern Belgium which are among the earliest regions anywhere to be industrialized, and which may therefore have been subject to contamination from long-obsolete industrial processes. In many of the older industrial sites, records of former land uses have been lost or were never fully kept. The EU member-states have, individually, attempted to cope with and respond to their past polluting activities. The unintended consequences or side-effects of their well-intentioned legislation now poses problems for the united Europe, especially for the operation of the single market.

Thus we turn to the issues that have arisen in the EU debates over the development of a coherent and consistent policy governing contaminated lands and their redevelopment. None of the questions we examine have been answered at the supra-national level — they still pose policy and legislative dilemmas for the Union. The variations in the national responses that emerged prior to the SEM remain as possible impediments to the rapid resolution of the policy dilemmas.

The EU Action — Subsidiarity Debate

The different national legal and financial approaches to contaminated land reclamation and liability risk certainly appear to constitute a distortion in the single market. The SEM, through the principle of freedom of movement of capital clearly includes the property development and investment sectors. Any variation in national cleanup standards and protection against liability for inherited pollution thus can have the effect of deterring investors from some member-states in favour of others.

There is a marked absence of empirical evidence on the extent to which this factor, among the many that may influence investment decisions within the SEM, affects the patterns of actual allocation of development finance. Vast differences, however, remain in practices regarding hazardous waste management, and trans-national issues arise from the cross-border shipment and disposal of potential land pollutants (Mangun 1988). Thus, because of environmental, as well as economic, concerns, there is growing pressure for an EU-wide response, notwithstanding the principle of subsidiarity.

It is, perhaps, ironic, that much of the argument favouring an EU policy is based on economic factors. *The Economist* (1994:90), for example, posed the question, 'might countries which adopt different standards on liability for past pollution thereby affect the competitiveness of their banks?'. German law, for example, permits banks to avoid acquiring property ownership and associated liability exposures in bankruptcy proceedings, while British law does not offer this protection. This difference in approach led the UK Department of the Environment to observe that, 'where a borrower defaults on a loan secured on a contaminated property, *British banks* are at a disadvantage compared with *European competitors*. In the UK they risk, when taking possession of the property, acquiring liabilities beyond the value of the advance; in Europe [sic], however, ownership is vested in the court, and only the advance and related sums are at risk' (UK DOE 1994:par. 4B.20, emphasis in original).

Another economic concern is evident in The Netherlands, which, with its high population density and substantial chemical industry, has concerns about pollution in residential areas. This problem has generated 'frequent mention of property devaluations [that] leads to a situation in which residents fear to be trapped....Their only way out seems to be either the full clean-up of the site or the necessity to leave their homes and move to another area' (van der Pligt and de Boer 1991). A localized land value shift may not be a matter for EU concern, but the quality and speed of cleanups and the impact on both population relocations and capital flows do constitute issues for the Union.

The European Parliament has exhibited concerns for contaminated lands that go well beyond their impacts on either individual property owners or on financial institutions and private redevelopment decisions. Commenting on the Commission's *Green Paper on the Urban Environment*, the European Parliament's Committee on the Environment, Public Health and Consumer Protection, for example, noted the opportunities for regeneration presented by derelict and contaminated lands in cities and stressed the joint economic and environmental benefits of coordinated approaches to such efforts. Most importantly, it recommended that the EU regional funds be utilized explicitly to promote reclamation of brownfield sites, even in regions not otherwise sufficiently economically depressed to be funding priorities (European Parliament (EP) 1991).

Resistance to EU action is strongest in the UK, despite its apparent disadvantage on liability grounds. The official UK position remains that subsidiarity applies and that it should be purely a matter for domestic legislation (UK DOE 1993). Yet, even within Britain, the position seems mixed. The Confederation of British Industries (1993) and the UK Environmental Law Association (1993), commenting on the EU *Green Paper on Remedying Environmental Damage* (CEC 1993a), both exhibit some ambivalence. The former supports the subsidiarity claims, but calls for the use of EU regional funds for remedying environmental damage, while the latter argues over different polluter pays concepts as it appears to praise the benefits of legal harmonization across the EU member-states.

Overall, however, the EU is an institution with a multiplicity of cultures and official languages, so the potential for common action may be undermined by problems of national definitions and terminology, with their implications for translation into common concepts. Interpreting legal and professional terminology in all the different official languages, and ensuring that the same understanding is conveyed to speakers of these languages, is a notoriously complex issue. Different interpretations of what constitutes contaminated land and how much land is affected is therefore both a terminological and linguistic problem, as the Commission's *Green Paper on Remedying Environmental Damage* (CEC 1993a) and the 1992 RETI report point out. Transfer of experience and promotion of a common understanding of the nature of contaminated land and of existing national policy instruments is made difficult by the regular generation of new vocabulary 'with a large dosage of words the original meaning of which has been changed to fit a specific use' (RETI 1992:37), the wide variety of disciplines (chemistry, regional development, environment policy, property development and so on) from which these words are taken, and the large number of words given quasi-legal meanings for specific purposes.

Some common European definitions are needed for an EU policy to be implemented. Both German and English, for example, use two terms for brownfield sites which have different implications. *Altlasten* and *Brachflächen* are both used in German to denote former industrial land in need of reclamation, but the former implies greater likelihood of contamination). Similarly, derelict land, and contaminated land in English connote abandonment in the first instance and pollution (whether the land is 'derelict' or not) in the latter. The term 'damage and impairment', not in use in the UK, is a Euro-English phrase devised by the Commission in order to try to achieve linguistic neutrality. As van Breugel et al. (1993) note, a simple glossary of terms is not enough to ensure mutual understanding given the cultural and legal system diversity across Europe.

The issue of level of intervention thus remains an open question. However, the subsidiarity argument against EU policy on contaminated land reclamation is likely to be challenged increasingly by other member-states motivated by the desire for higher overall standards, removal of distortions of competition caused by the uneven spatial distribution of contamination, or the expectation that assistance from an EU regional fund or other source may follow from the adoption of an EU Directive. The increasing attention given polluted sites in EU discussions of other supra-national concerns, such as their inclusion in the Commission's 1993 *White Paper on Growth, Competition and Employment* (CEC 1993b) suggests that formal EU action on the contaminated land problem is virtually inevitable.

Cleanup Standards

Despite over a decade's experience under CERCLA, the US has yet to arrive at common and consistent standards for site mitigation. The question of 'how clean is clean?' presents even more problems in the EU context, since individual member-states have their own standards in place, and revising them presents problems such as the need to re-initiate cleanups on sites not meeting possible higher reclamation criteria. The basic conflict is between mitigations suitable for intended uses and multifunctional restoration.

The UK adheres to a 'suitable-for-use' approach to the question of deciding the extent to which contaminated land needs to be treated. The object is to limit the expenditure by not imposing a requirement to improve land beyond the standard necessary for any proposed future land use (UK DOE 1994, paras 2.5, 2.6). The alternative is to restore land to its original condition (which would be very expensive) or on the basis of any potential future use (the multifunctional approach already applicable in The Netherlands or Germany). The suitable-for-use approach is inherently short term, despite its lower cost, presuming rigidity and certainty in future land-use plans. The implied limitation on any future land uses is deemed unrealistic in Germany, which is why, despite its massive contamination problems, it opted for multifunctionality (Genske and Noll 1994).

The differences in standards that now exist across the EU raise the potential for a distortion of the single market. We have found no evidence of their impacts on capital flows and regeneration rates. Nonetheless, arriving at a common standard, which is what the rejection of subsidiarity implies, presents major policy and political difficulties.

Attribution of Liability for Damage

The normal expectation under property laws in Europe would be that the current owner would, when acquiring the bundle of rights and responsibilities that collectively constitute ownership, have acquired liability for any contamination on the site (under *caveat emptor*) unless a previous owner can be identified as the polluter. The nuances of implementation of this principle, however, vary substantially, especially with respect to environmental damage caused.

The variation in the treatment of liability across the EU member-states, has been documented in a report on four countries conducted for the UK Scottish Office (Connell 1994). Connell found 'neighbour laws', analogous to the nuisance provisions of common law in Denmark, France and Germany, all providing for some imposition of strict liability. The Netherlands, while not providing for strict liability, is exceptionally stringent in its interpretation of fault-based liability, generally demanding proof from parties charged with damage that they were *not* the cause of the problem. All four countries, under a variety of provisions, paralleled the UK in application of negligence standards, especially for any damage resulting from actions in violation of the laws or standards prevailing at the time they took place. While generally appearing to assign strict liability more than the other countries, however, Germany was the only one to legislate a maximum liability exposure (160 million Deutschmarks; roughly $90 million). Italy and Portugal have both passed legislation providing strict liability for current 'dangerous activities in general', but how these provisions may apply to past damage is not clear (CEC 1993a:14).

The European Commission's 1993 *Green Paper on Remedying Environmental Damage* placed the question of adoption of the US CERCLA principle of strict, joint and several liability for past contamination on the EU agenda. Opposition to the application of this principle is strong in the UK, with the Confederation of British Industries (1993) and the UK Environmental Law Association (1993) in full accord with Her Majesty's Government (UK DOE 1993) in rejecting both strict liability for contamination and the joint and several principles. Their position is that current owners should be fully liable for the conditions of their holdings. The situation is more fluid and less unanimous on the Continent, according to Connell (1994): while Denmark is moving away from its prior tendencies to hold non-polluting owners liable for past contamination on their lands, both Germany and The Netherlands are moving towards the expediency of such a principle; France attempts to attach liability to specific polluters, not those with land title.

Differences in national approaches to liability attribution may impede the development of an EU policy. The difficulties in arriving at a consensus are starkly evident in such current trends as the movement in opposite directions of the Danish, and of the German and Dutch, approaches to current owner liability for past pollution. Nonetheless, the pervasive presence of legislation assigning *strict liability* for environmental damage in the member-states (CEC 1993a:Annex 1) suggests the plausibility of using this legal principle at the EU level.

Rationales for New European Funds

The regional funds administered by the European Union exist to provide greater equality across the broad variety of economic conditions that exist between different member-states and sub-national regions. The argument that analogous funds should exist to provide an environmental level playing field are persuasive in their reliance on precedent. Both the Green Papers we have examined, that on the urban environment (CEC 1990) and on remedying environmental damage (CEC 1993a), as well as the proposed Directive on liability for waste (CEC 1991) suggest new EU funding mechanisms.

However, two issues need to be addressed before any decision could be made on a new European Fund. Different types of costs are involved in any reclamation efforts, as we noted in Chapter 1; for each of the two broad classes of cost sources, cleanup charges and liability risks, the SEM and its need to provide a level playing field pose different questions about appropriate allocation of costs between the public and private sectors.

The liability issue has dominated debate in the UK (as in the US), but the cleanup costs themselves have been the major concerns in most other EU countries, despite the spread of new strict environmental liability provisions and the continuing modification of environmental liability laws, which we have just examined. The difference between the UK and its EU partners in this regard may, in part, be attributable to the prevalence of a corporatist ethos on the continent and the UK adherence to (or drift towards) the individualism and free market ideology that is characteristic of the US. There is less difference between the member-states on the principles guiding private and public responsibility for cleanup.

Contaminated land reclamation is, in the first instance, the responsibility of the polluters — or current landowners wishing to use the properties for new purposes (that is, engaging in redevelopment). This proposition is broadly shared across the member-states, although there is variation in the state provision of funds for mitigation where no owner or past polluter takes responsibility for cleanup. The UK's derelict land grants, initiated in 1966, is arguably the first provision for abandoned lands, although the programme has grown to cover more and more redundant state-owned property, so the grants really represent application of the polluter or owner pays principle, with the state as owner. The Netherlands and Denmark both have state-financed cleanups, with provisions for cost recovery from specific identifiable polluters. Sweden also conducts state cleanups where private parties cannot be charged with the duty, but recovers costs through a tax on all installations requiring current environmental permits. (The Danes and the Dutch thus parallel CERCLA cost recoveries; Sweden's financing its fund through taxation of potential polluters recalls the Superfund industry taxation provisions. The difference is a stricter 'clean now, collect later' policy in the European nations.) Germany's Treuhand agency has taken over cleanups in the former GDR; the national state is the logical successor to the state enterprises of the communist East that produced massive pollution.

Cleanups clearly are not the exclusive responsibility of the private sector: the state has played a role across Europe (as in the US). Thus the question of an EU fund to assist reclamation and cleanup of past contamination is one of what the criteria for support eligibility may be. When the polluting entity exists and fault for the contamination can be attributed, it can and should be held accountable for cleanup. If the polluter cannot be identified, no longer exists, or is not solvent, there are clear grounds for state intervention — or EU fund eligibility.

Questions arise if the costs of cleanup might bankrupt an otherwise viable entity or in some other manner pose an 'unreasonable' burden. Precisely analogous issues arise for liability for past actions, since such costs, too, could cause an entity to cease to exist and make an economic contribution. The response to such questions should, logically, depend on the regional, national or EU impacts of economic damage to the polluting or liable entity.

An equity consideration also arises, however, for any fault-based cleanup responsibility: should current damage be done to an economic unit for past actions that were legal and acceptable when they were undertaken? The same question might be asked about fault-based liability for damage caused by past pollution, since the member-state emphasis on strict liability we have noted applies primarily to current, not past, operations. These issues remain problematic for decision-makers in the member-states as well as for the Union as a whole; the conundrums themselves are not, however, an adequate basis for rejecting EU action.

Either high liability exposures or cleanup costs associated with reclamation of particular brownfield sites may retard cleanups and redevelopment, provided other investment options exist for mobile capital. Capital mobility unquestionably has increased under the SEM, so the 'level playing field' argument supports some EU regional funds for environmental purposes. The pervasive member-state practice of subsidizing brownfield development relative to greenfield options within limited geographic areas suggests an awareness of the investment diversion problem, and may presage eventual EU action on new environmental reclamation funds.

DIVERSITY IN MEMBER-STATES' CONDITIONS AND POLICIES

Whether or not a consensus — or QMV majority — can be attained on the promulgation of an EU contaminated land policy and formal implementing Directives may depend on single market considerations. The form and content of the Directives, however, may be constrained and shaped by the need to integrate the existing diversity of national land-use planning policy and past pollution experience of the member-states. Depending on their historical evolution, environment and land-use planning codes may be closely integrated or quite separate bodies of legislation. We examine here five factors that shape national problems and perceptions of the issues associated with reclamation of contaminated land. Then we turn to two examples of national policies, the very proactive German approach and the market-oriented UK practice, and case studies of their impacts.

1. The structure of member-state government, locus of environmental responsibilities and powers, and degree of federalism, especially as regards land-use planning, affect policy implementation. The extent to which powers to legislate as well as to implement or enforce policy are held at the regional or upper tier of sub-national government, as well as the controls exercised at the local or municipal level are additional key variables. In Belgium, for example, land-use planning legislation is now entirely a matter for the three regional governments (Laconte 1992) as it was until 1959 in Federal Germany. The 1986 Federal Building Law Code (BauGesetzBuch) in Germany sets a framework for any state legislation (Dieterich et al. 1993). The *Länder* have more authority in shaping land-use policies than do the French *Départements* that are sub-units of the national state, but less than do the Belgian regional governments.

2. Two different pure types of controls may be instituted to protect the environment, environmental quality (or impact) standards and

emission (or action) standards. The former is, arguably, less intrusive as regards private firms' activities, but may, in effect, more directly alter environmental conditions. While the UK leans more towards the former than the other EU member-states, all involve some mix of the two approaches, as do the existing EU Directives on environmental policy (Haigh 1989).

3. The rigidity of national controls, paralleling the land-use planning systems to which we have alluded, affects the effectiveness of controls. Roman law countries tend to maintain legally-binding plans (Davies 1989; Dransfeld and Voss 1993); the addition of the Code Napoléon to the Roman law leads to the most rigidly codified approaches (e.g., France, Belgium), but an overlay of common law, as exists in Sweden, induces more flexibility (Wood and Williams (eds) 1992; Acosta and Renard 1993; Kalbro and Mattsson 1995). The development plans and environmental designations that would be adopted in accordance with the (common) law in the UK would, in Germany, *be* the (Roman) law. The difference is critical to appeals of plans, which are strictly judicial where the Roman law prevails.

4. Differences in national political cultures and policy-making styles still prevail and cannot be ignored. What may appear to some as rather dubious forms of patronage, influence or behind-the-scenes negotiations may be accepted as perfectly normal elsewhere. This is especially problematic for development planning that is actually implemented at a local level, where interpersonal ties are more likely, rather than directed on the national stage.

5. National cleanup imperatives may be as much a function of geographic factors and industrial history and geography as of expressed political will. The fact that Lisbon is a city with clean air, while Athens is notorious for air pollution, for example, is more a product of topography and wind movement than of technological or attitudinal differences.

Land Reclamation and Redevelopment in Germany

German public policy has emphasized the achievement and maintenance of high environmental standards. This is most evident in the reclamation efforts in the Ruhrgebiet in Nordrhein-Westfalen,

Germany's oldest and largest heavy industry region that contains 71.3 per cent of all derelict land in the former Federal Republic (pre-reunification boundaries).

Public–private partnership agencies (PPP), with the public sector playing the lead role, have been central to these efforts. The LandesEntwicklungsGesellschaft (LEG — an economic regeneration agency for the region established by the state government), the MontanGesellschaft and the EntwicklungsAgentur Östliches Ruhrgebiet (EWA), both engaged in restoring and recycling land for new development from former steel plants and coal mines, have played key roles. Investment is encouraged and must normally be subsidized, either directly or through removal of risk of liability from developers and investors. The public partners decide which projects or sites to address, take responsibility for the cleanup operations, and arrange for the development and occupancy permissions required. They thus ensure that the private partners (developers) are not at risk from inherited liabilities for pollution. Participating in marketing the sites, they then use any ensuing profits to invest in their next projects.

In 1980 a special fund, the Grundstücksfonds Ruhr (Ruhr site fund), was created by the state government, which entrusted the LEG with its operation. The LEG has used the fund to temporarily take title to lands previously held by such large landowners as mining companies, steel works and German Federal Railways, which between them held most of the Ruhr's contaminated sites. In so doing, the LEG or other PPPs overcame the owners' reluctance to sell by offering prices up to owners' book value and by accepting state responsibility for any remaining liability for waste and contamination (Pohl and Grüssen 1991).

Public awareness of the contaminated land issue was heightened in the mid-1980s by the discovery of unusually high rates of illness among residents of a housing project in Dortmund on the site of a former coking plant. The discovery of substantial site contamination generated public pressure and led to the municipality purchasing the houses and cleaning up the site. This case (which is remarkably similar to the Love Canal incident in the US) has led to tightening of procedures to test for contamination before any planning and building authorization is given, and to considerable research effort on improved recycling and reclamation techniques (Pohl and Grüssen 1991).

This experience may have led to the 1989 foundation of the Abfallentsorgungs- und Altlastensanierungsverband or AAV (waste and contaminated ground disposal association), which was given the legal authority to issue licences to private waste disposal agents to clean up 'orphan' sites without exposure to legal responsibility for the contamination. The AAV protects the capacity of past polluters to remain in business despite their contamination debts, while facilitating abandoned site reclamation through its capacity to provide between 70 and 90 per cent of the finance needed to ensure decontamination and removal of health hazards (AAV 1993).

The results of these efforts at state–private sector collaboration are evident in a series of major new projects and initiatives across the region. Düsseldorf, just west of the Ruhrgebiet, has benefited from the programmes indirectly. As a major administrative and commercial centre, the city has been able to attract investors in spite of its contaminated sites, since property values permit profitable development even when the private sector bears all the costs of reclamation. However, the Ruhr's PPP precedents have been important in stimulating agreements to protect developers from future pollution damage liability claims, as the Hansa Park office development on an abandoned Thyssen steel company site in central Düsseldorf that now provides jobs for over 1,500 workers demonstrates (Dieterich, Dransfeld and Voss 1993).

The need for public subsidy for reclamation of brownfield sites in the Ruhr is well illustrated in the case of the Dortmund Technologiepark. The city, at the eastern end of the Ruhrgebiet, has been relatively successful in overcoming the loss of its traditional heavy industry employment base and in creating a positive environmental and cultural image. However, promotion and development of the new university-linked science park necessitated a subsidy of DM 15.5 million despite its location on a *greenfield* site of 19.5 hectares adjacent to the university in one of Dortmund's clean air corridors protected by the regional plan (Dieterich, Dransfeld and Voss 1993). In a setting in which even the sacrifice of pristine and preserved land is not sufficient to attract investment capital, the negative effects of a poor local image and the remaining barriers to brownfield reuse are clearly massive.

An approach used in Germany and elsewhere in Europe to address the image issue is the visionary international exhibition or grand design for an area or locality as a framework within which specific direct action takes place. The idea of an Internationale Baustellung (IBA, or International Building Exhibition) has a long history in Germany, usually with an architectural purpose. In the case of the Emscher Zone of the Ruhr, an area with some of the most acutely contaminated land, the IBA concept was employed to transform the area's image, stimulate international attention and generate new investment in reclamation and regeneration (Hennings and Kunzmann 1993).

Bitterfeld, Sachsen-Anhalt, in the former German Democratic Republic, an area of former lignite mining and chemical industry with some of the worst contamination in the EU, is another case where a visionary concept has been adopted in order to replace its negative image with a new and positive one as a region offering new investment opportunities. The programme there goes by the title Industrielles Gartenreich (Industrial Garden Realm). To the pragmatic Anglo-Saxon mentality this type of concept appears as mere rhetoric, but it would be wrong to conclude that it is of no value in attracting developer interest and overcoming the image problems associated with contamination.

Bitterfeld was the location of one of the largest chemical complexes in Europe, characterized by out-of-date technology, minimal maintenance, and air that was unsafe to breathe, largely due to daily accidental leaks. Ongoing restoration efforts include clearance of buildings and plant, itself a hazardous operation, and removal of up to three metres of topsoil from the site of the chemical complex itself, at a cost of over DM 2 billion. An additional DM 12 billion programme to remove contaminated material and restore over 10,000 hectares of former open-cast lignite mines was announced in December 1994.

The air is now much cleaner and, importantly for the local economy, many jobs are being created not only in the reclamation programme but also in a new chemical-based industrial park located on part of the contaminated former industrial area, the Chempark Bitterfeld-Wolfen. Investment in and business occupation of sites at this new manufacturing centre was stimulated by the new image promoted by the designation of the Industrial Garden Realm by the Bauhaus, Dessau (Toyka 1993).

Land Reclamation and Redevelopment in the UK

The UK has actively addressed the specific problem of contaminated land approaches since 1990, with two underlying leitmotivs: acceptance of the US experience as a model to inform policy-making in the UK, and rejection of the notion that there can be any role for the EU on legislating for contaminated land and associated liabilities. The institutional, cultural and geographic differences between the US and the UK appear to have been ignored in the government's interpretation of US experience, while the insistence on subsidiarity has precluded consideration of the possible benefits of an EU policy.

Financial assistance to overcome problems of abandoned and potentially polluted land has, as we have noted, long been available nationally under the Derelict Land Grant (DLG). For land recovery schemes that qualify, DLG has provided 100 per cent of the eligible costs incurred by local authorities, repayable if the authority realizes a profit on its land transaction after cleanup and site preparation. Private firms are eligible for 80 per cent grant, but with stricter conditions concerning possible repayment. Grants are also available under other regeneration programmes, as the DLG system has been incorporated within urban policy. The DLG itself does not protect its recipients from any risks of lingering pollution or associated liability.

The DLG has also played a role in the closest parallel in UK policy to the IBA–Emscher concept, the cycle of National Garden Festivals in 1984–92, an image-changing approach actually adopted from the well-established German tradition of Garden Festivals. Several of the sites selected in the UK, such as Glasgow in 1988 and Gateshead in 1990, won major funding in a competition whose criteria included the presence of acute local contamination as well as imaginative festival plans. Cities gaining such festivals have been funded for derelict land reclamation in one or two preparatory years at levels equivalent to what would have been possible over thirty years at their prior level of DLG support.

Yet the impact of the DLG overall has been limited. A 1992 EU environmental assessment observed that the UK had 43,000 hectares of industrial land in 1994. While 2,000 hectares have been cleaned annually since that date, another 2,500 were abandoned (CEC 1992b:42). As a result, a 1990 House of Commons Select Committee on the Environment report estimated that as many as 100,000 UK sites could still be contaminated. The problem of land contamination has been growing, not shrinking.

The 1990 Environmental Protection Act is, arguably, a legislative watershed. It criminalized the act of pollution for the first time and introduced the concept of 'duty of care', a legal obligation on all parties involved in handling any controlled — potentially toxic — wastes, akin to that created in 1976 in the US under RCRA (Harris 1992). However, the element of the Act that has proved most controversial is Section 143, which required the creation of public registers of contaminated land by local planning authorities. To minimize site assessment costs and the time required to create the registry, it was to include all sites found to have been used for purposes that potentially could have caused contamination (requiring for each site roughly the equivalent of the US CERCLA Phase I site assessment).

This broad inclusion led many to argue that it would have an unnecessarily wide blighting effect, asserting that listing would bring land stigmatization, especially since the register would also serve as a record of cleanups and sites would never be de-listed (Chakravorty 1992). The possible gains from reducing the costs of a *caveat emptor* property market did not receive much attention. In response to protests from both the development industry and local authorities concerned about economic development, the government suspended preparation of the contaminated land registers in March 1993.

Later that year, a House of Lords Judicial Committee's judgment on a court case (Cambridge Water Company vs Eastern Counties Leather plc) limited strict liability for past pollution, ruling that the plaintiff did have to establish that the specific damage was reasonably foreseeable at the time the contamination took place (UK DOE 1994). This decision has been hailed by past polluters and those concerned with retrospective liability for past contamination, since it may have effectively eradicated such strict liability exposure. Environmentalists have, by contrast claimed the ruling turns the polluter pays principle into an impediment to environmental litigation (Shelbourn 1994).

The UK Department of the Environment responded in March 1994, with a consultation paper on contaminated land which reopened many of the issues, including both the issue of the land registries and the question of imposition of strict liability (UK DOE and Welsh Office 1994). The document, *Paying for our Past*, elicited responses from the planning, economic and development communities, and, after completion of the consultation process, was followed in November 1994 by announcement of the policy the government now intends to pursue, *Framework for Contaminated Land* (UK DOE 1994). It proposes legislation to strengthen the national framework for remedial and regulatory action, giving powers to a new Environment Agency and to municipal authorities. Key points are: (1) repeal of S.143 of the 1990 Act; (2) no promulgation of absolute cleanup standards; (3) rejection of a 'deep-pockets' attitude to financial institutions' liability (akin to that under CERCLA); (4) no required remedial action by owners or the state except under conditions of unacceptable risks to health or the environment related to actual or intended land use. No further legislative definition of liability is proposed, as the government now accepts the view that the judgment in the Cambridge Water case achieves a satisfactory balance between strict and fault-based liability within Common Law while adding useful flexibility (UK DOE 1994).

The fourth provision is the 'suitable-for-use' principle which has the effect of setting lower minimum standards than those of Germany, The Netherlands and even the US. However, it carries the further possibility of setting some contaminated sites aside for delayed action. The UK thus, in effect, adopted some of the worst features of CERCLA without its stronger provisions: no assurance that sites will be cleaned, no access to past polluters or deep pockets to augment limited public funds for cleanups, and no effort to arrive at consistent cleanup standards. It has certainly demonstrated an intent to head in a very different direction from its fellow EU member-states, a path that may not be sustainable, given the new EU balance of power.

Diversity and the Potential for EU Policy-Making

Contaminated land does appear to be an issue for European Union action and legislation both for single market reasons (Williams 1995a), as even the UK DOE paper in effect acknowledges (UK DOE 1994), and because of commitments to a high environmental standard and to the precautionary principle embodied in the Environment Title of the Treaty of European Union and the Fifth Action Programme on the Environment (CEC 1992a). The lack of EU attention to this issue despite the fact that it involves tens of thousands of hectares may be explained by Germany's focus on the technical issues of site mitigation, the anti-Europe posture of the ruling Conservative Party in the UK that has led to blocking efforts, and the simple fact that air- and water-pollution issues are more prominent than land contamination in most of the remaining member-states.

There is still no detailed evidence of the impact of different policies and legal frameworks on the behaviour of individual developers and investors. Some finding of a serious market distortion is a precondition for EU action on market grounds alone. Without such information, the issue is less likely to attract the Commission's attention; even if it did, no design of an optimal policy is possible, albeit directions for change may be discernible.

PART IV

7. Comparisons and Contrasts: Integrated Comparative Analysis

We now turn to some comparisons of the contexts in which contaminated land policies have evolved in the nations of Europe and the United States. As we consider the similarities and differences in experiences, we take as our initial comparative framework the EU relative to the US federal government as the central regulatory body, and the EU member-states and US states as 'local' polities, responding to central mandates, standards and policy directions. We will turn later to the comparison of the US as a member of a free trade zone, NAFTA, to the EU member-states operating under the Single European Market.

Having laid out in Chapter 2 a series of dimensions that shape contaminated land policy, we return to that framework to distil findings from our review of policies and experiences in the US and the EU. We address each of the major categories discussed in Chapter 2 in order, using a tabular framework to lay out the comparisons for the more complex sets of dimensions. Once we can summarize major contextual variations, we then turn to extracting evidence on the development cost implications of the approaches in the EU and US, as they appear to have evolved to date. For this purpose, we return to the arrays of development transaction costs and impacts on returns to investment in land presented in Chapter 1.

PAST ENVIRONMENTAL EXPERIENCES: ACCIDENTS, EXPOSURES AND 'DISASTERS'

Whether or not they are the direct cause for the passage of new laws and regulations addressing issues of hazardous waste and land contamination, accidents and public exposure to toxins appear to have stimulated political concerns and generated demands for action on both sides of the Atlantic. While CERCLA was drafted by the EPA in the United States prior to the discovery of dioxin at Love Canal, the publicity surrounding the evacuation of the local residents certainly helped move the legislation forward. Similarly, in Europe, it is not coincidental that the 1982 EU *Directive on the major accident hazards of certain industrial activities* is known generally as the 'Seveso Directive', after the 1976 accident that sprayed residents of that Italian city with dioxins.

Such events help galvanize interest in the complex problems presented by an economy's routine use of highly toxic substances in production and the public health problems their use, transport and disposal may present. Well-organized business interest groups have strong profit incentives for resisting intrusive requirements for private sector handling of toxics releases and reporting on their release into the air, water or land. The 'public interest' is not generally as well organized to wield political influence. To some degree, then, it is probable that, in the absence of well-publicized accidents or exposures, the environmental laws and regulations promulgated in the US and EU would not be as stringent as they now are. Certainly Barnett has strong grounds for his claim that the 1984 accident at Bhopal, India 'catapulted a right-to-know title to a position of prominence on the redesign agenda' as the US Superfund reauthorization bill that emerged as SARA wound its way through Congress in 1985 and 1986 (1994:196). The same causal chain may apply to the 1987 amendments that significantly broadened and strengthened the original Seveso Directive.

We cannot use the experience of accidents or their absence to distinguish differences between EU and US policy, since both jurisdictions routinely encounter local accidents to which the public reacts — and both are exposed to news about distant accidents, whether on the opposite side of the Atlantic, such as repeated news about oil spills, or on the other side of the globe, as in Bhopal. The context in which policy emerges in both the EU and US is well summarized in an OECD report on accidents:

> In general, hazardous substances are produced, transported and used under safe conditions in OECD Member countries. These activities nevertheless result in several hundred accidents a year, causing significant damage to public health, the environment and property. In some cases, damage to health and the environment has extended far beyond the immediate site of the accident, and on occasion has taken on catastrophic proportions. (OECD 1989:7)

While massive loss of life has not been directly linked to even the occasional event of 'catastrophic proportions' in the OECD member countries, the fear of such disasters is real and ever-present, and constitutes a critical component of the political environment in which approaches to contaminated land are shaped.

THE STATE POLITICO-LEGAL CONTEXT

The role of the state and relations between the central jurisdiction and its component units vary substantially between the European Union and the United States. Table 7.1 summarizes our findings in reviewing the situation in the two settings. The essential differences lie in the power of the central body and the means by which it directs policy implementation, in the rights of local units to modify the central Directives, and in the approach to standard setting for cleanup of contaminated land. All the major differences evident seem to be driven in large part by the major divide evident in the reliance on the state relative to markets, in other words, the extent of the corporatist or individualist ethos guiding public policy.

The EU, overall, is far more corporatist than the US, with the majority of its member-states accepting primary responsibility for cleanups, sometimes even suppressing the possibility of independent private development efforts. The strongest individualism in Europe is exhibited by the UK, which has been adamant in exercising its subsidiarity rights to limit the intrusion of strict EU policies into its land markets and encouraging publicly-created entities such as the Urban Development Corporations to speculate on returns to mitigation of polluted sites.

While Germany, Italy and The Netherlands use area-wide approaches to contamination, the UK has relied far more than its neighbours on the site-specific approach employed virtually exclusively in the US. Areas are not marketable units of land, unlike sites, so site-specificity is logically consistent with an emphasis on market processes and relied upon in the US despite the high probability that an area-wide approach would be more efficient, especially in terms of reversing perceptions of dereliction and contamination. (The Wichita case, which does exhibit an area-wide policy, may be explained as a preventive response to the imminent collapse of local real estate market processes.)

The rigidity of the federal policy in the US, dictating both processes and standards to the fifty states and local governments involved in regulating and reclaiming contaminated land, may seem perverse in an

Table 7.1 The state politico-legal context

Dimension	European Union	United States
Mandates from central jurisdictions	Varies; defines legal and policy objectives, and, at times, sets exposure standards	Substantive, setting standards and requiring specific outcomes
Reliance on the state and markets	State accepts primary cleanup responsibility as reflection of varying degrees of corporatism	Markets primary, with state as mitigator of last resort, reflecting an individualist ethos
'Local' options and flexibility	Subsidiarity, with local flexibility	Central dominance; local subordination
Spatial policy approaches	Local areas, regions	Site specific
Standards for cleanup	Variable across member-states; e.g.: UK — based on intended use; The Netherlands — multifunctional	Uniform, to a fixed cleanliness level

individualist context. It appears, however, to reflect a distrust of more local jurisdictions that compete for new economic activity by providing an array of implicit and explicit subsidies to private firms. Robert Goodman's (1979) characterization of the states as *The Last Entrepreneurs* for their activities in pursuit of new businesses and jobs remains true today in the US, for all that their ranks have been joined by counties and cities prepared to take on almost any burdens to attract new capital and economic activity. Given the well-developed 'market' for local deregulation as a commodity, a form of subsidy offered in return for the in-migration of mobile capital, the federal government in the US has little choice but to dictate minimum standards for process and outcome if sub-national competitive processes are not to expose the public and ecosystems to risks from contamination it deems unacceptable.

THE LOCUS OF CONTROL OVER THE USE OF LAND

The differences between the EU and US in Table 7.2 continue to reflect the corporatist–individualist distinction between the two settings. The nature of control over land use, to a greater extent than the more general characteristics of the state and its powers, has specific implications for the impact of land contamination on the prospects for urban regeneration. The generally high levels of power to control the use of land and expansion of urbanized areas in the member-states of the EU combine to severely restrict the options available for would-be developers, relative to those available in the US. As a result, firms that pursue profits through land development activity have fewer options available and are more inclined to invest in brownfield sites. The limits on urban sprawl and land-use alternatives, moreover, may raise urban land values in the European nations, independent of their generally higher population densities, promising greater returns on reclamation and redevelopment projects.

The capacity of the EU member-states to acquire land for any purpose facilitates state action to mitigate contamination and prepare brownfield sites for private development. By contrast, the US practices not only constrain the possibility of state-led reclamation of privately owned contaminated sites, but render the implementation of preservation and development limits more difficult, and recent court decisions increase state obligations to compensate landowners for any restrictions imposed on permitted uses of their lands that are promulgated after they have taken title (cf. Franzen 1994; Tibbetts 1995). Given its legal traditions and court decisions on property rights, the US, in most instances, has to employ regulatory controls on private parties, and thus relies on the threat of punitive damages (the treble damage provision under CERCLA) to produce 'voluntary' cleanups by potentially responsible parties.

The broad rights of public access to private lands in most of Europe, not present in the United States, tend to drive up the requisite standards for mitigation of contamination, even in EU member-states with standards that vary with intended use. Unless that use vests landowners with rights to exclude the public, the problem of casual user exposure to residual toxins in the case of weak cleanups could present a problem, both for public health and for the potential liabilities of the landowners.

Overall, the various elements of the locus of control over the use of land combine to produce more incentives for private redevelopment in Europe than the US, as well as a greater capacity for the state to conduct reclamations itself. Other things such as access to capital and exposure to liability being equal, the negative local economic development effects of land contamination should tend to be weaker in European nations than in the US. The UK and Ireland, in which zoning provisions do not have the rigidity of the continental European states, are the most likely to encounter problems akin to those of the US. Their smaller size and higher density of settlement, however, combined with their utilization of greenbelts around major cities, generates a level of land scarcity that is not present in the US, and, with it, greater incentives for private sector-led brownfield reclamation effort.

APPROACHES TO LIABILITY FOR DAMAGES AND CLEANUP

The treatment of liability varies substantially across the member-states of the EU. The responses of the different US states to the economic development problems they perceive in CERCLA has led them to a similar range of prospective liability exposures, but none have risked taking on full retroactive liabilities for current contamination. Table 7.3 describes the governing approaches of the nation-states, so it does not do justice to this variability within the US itself.

Table 7.2 The locus of control over the use of land

Dimension	European Union	United States
Exclusivity of property rights	Low: public access provision is high and ownership prerogatives limited	High: limited state controls of owners; litigation over private ownership rights
Zoning approaches	Generally rigid and controlled	Minimally controlled
Legal immutability	Varies between countries with Roman law traditions that are rigid and the English common law states that are more mutable	Highly mutable with frequent discretionary zoning variances
Zoning logic	Predominantly exclusionary	Predominantly hierarchical
Land-use controls	Active permissions to act required, yielding high levels of control	Presumptive permissions to act and limited control
State powers to take land or expropriate	Compulsory purchases or expropriation for *any* state purposes, even if land is returned to private use immediately	Eminent domain for public facilities only; compensation to owners may be due for loss of potential profits on land speculation
Preservation and development limits	High; legally mandated and supported	Low; limited legal mandates
Architectural constraints	Listed buildings	Listed buildings
Preservation of open space	Planning controls, urban containment policy, green belts and protected zones	No general powers to control urban sprawl

Table 7.3 Approaches to liability for damages and cleanup

Dimension	European Union	United States
Environmental damage liability	Current owners and users; state acceptance of responsibility varies	Shared across all past and current users and owners
Cleanup liability	State primacy, with contributions from current owners, users	Past and current users and owners; state enforcement using industry-financed Superfund moneys
Liability allocation across private parties	Identifiable past polluters only; state responsibility for past polluters that no longer exist	Strict, joint and several for all parties, potentially including lenders
Treatment of new landowners	Protected from liability	Unprotected from liability
Required disclosure	Some disclosure required; UK also relies on *caveat emptor*	US level: *caveat emptor* with no required disclosure; some state disclosure requirements
New purchasers' liability shares	None, unless knowingly accepted	Potentially full regardless of fault
Treatment of future liability	Not private, but state responsibility	Fully assigned to private landowners
Public–private sharing	Primarily a state responsibility	No limit on private liability
Private sector cost sharing	Any private burden is on past polluters, not current users	Current users liable, regardless of fault
Public land ownership	High	Low
Total public landownership	High, especially for land used in production	Low, except for parks, preserves and military bases
Sub-national urban public landownership	High; limited number of entities	Low; large number of public entities

In nation-states with high levels of state landholding, there may be little difference between the 'landowner' and the state when it comes to the realities of who pays for damage done by past contamination and for cleanups. This is especially true for the many EU member-states that have had nationalized heavy industry in the post-war period. All bear state responsibility for contaminated sites on or near railway lines, with all rail service nationalized, and most of them have operated steel mills, automobile manufacturing or other such major polluting manufacturing installations.

The US, on the other hand, not only has a higher proportion of land in private hands, but that land that is held publicly is held by a much larger number of sub-national state entities than is the norm in Europe. Any principles of liability that tend to particularize responsibility in landowners will, in such a setting, tend to highlight the differences between the interests of particularist local state units and the national state. That is, a small public sector unit, whether single-purpose or not, acts in the interests of a small proportion of the total national population. The issue of whether or not the citizenry at large should bear the responsibility or the costs for public sector action by such a unit becomes analogous to the same question for actions by private landowners or users.

The high level of public landholding in the EU member-states may thus help explain the tendency of the state to play a primary role in assuring mitigation of contamination in most of the countries of Europe. If all taxpayers share responsibility for past pollution by state-owned enterprises and such pollution comprises a major proportion of the total contamination, it is not worth the administrative effort of determining whether the enterprise in particular or the general state budget is to be billed for the relevant costs. If the costs are allocated to the enterprise, it will probably have to have its state subsidies raised to cover the costs, and will produce the same charge against general revenues as would exist if the state accepted primary responsibility for correcting the contaminated condition.

The extent of public landownership, however, does not explain the differences in private liability allocation, either in the treatment of future prospective liabilities, or in the treatment of new landowners. The variation in all three of these aspects of assigning liability appear to be associated with some other factors.

The treatment of new landowners does appear to be linked to the extent to which a corporatist or statist orientation governs policy formation and implementation. From the UK similarity to the US, it appears to derive from the extent of reliance on the market we have already discussed: any deviation from adherence to the *caveat emptor* principle constitutes a state intrusion into market processes. In these two most individualist cases, the least protection is provided to 'innocent' new landowners. Significant variation between the UK and the US does exist, however, and the individual US states themselves exhibit very different patterns, as we have seen.

New Jersey, for example, followed the 1980 passage of CERCLA with its own legislation in 1983 that not only required disclosure of contamination as a condition of sale, but further demanded that a cleanup plan be in place before a property transfer could take place. Other states have followed with disclosure requirements, including Michigan and Illinois, but they have not imposed the cleanup conditions. These state interventions in their property markets appear to have been motivated as much by the need to reduce buyer uncertainties and risk exposures in order to attract development capital as by the desire to locate and treat contaminated lands.

While many US municipalities are conducting site assessments prior to taking title to tax delinquent lands (in order to protect themselves from liability), they have not, by and large, disclosed the extent of the contamination identified. Milwaukee, Wisconsin, and Atlanta, Georgia, have made their lists of contaminated tax-delinquent properties public, but other cities, such as Louisville, Kentucky, have avoided disclosing the known extent of their stockpile of contaminated lands. The US cities appear divided in their approach to the questions raised in the UK over implementation of Section 143 of the Environmental Protection Act 1990 that required local authorities to compile and maintain a register of lands they considered to be at least potentially contaminated. The differences in approach they exhibit may reflect their local land-scarcity levels: Milwaukee and Atlanta are both much larger than Louisville, and are centres of much more intensively developed urban areas, so they have a relatively greater scarcity of alternative developable land — and the expected returns on reclamation of comparably risky contaminated sites is accordingly higher in those settings.

The very different approaches to liability allocation across private parties, however, do not appear to have roots either in a cultural/political ethos or in particular property market conditions or processes. The nations of Europe, by and large, rely on fault-based liability while the US has chosen to impose strict liability on all *potentially* responsible parties (PRPs), regardless of blame. We have

Comparisons and Contrasts: Integrated Comparative Analysis 157

already seen that the latter approach has caused massive administrative and litigational problems that have had the net effect of retarding cleanups in the US. The apparent rationale for the US approach, however, was its administrative *simplicity*! The argument was that the EPA should not be burdened with the problem of allocating cost shares, and the use of strict joint and several liability permitted the Agency to simply name a single party as responsible for a site cleanup, presumably one with the financial resources to complete the task.

In the ideal, the Agency would have left the many PRPs to fight out the issue of cost allocation among themselves while one of their number executed the necessary mitigation of the contamination. Obviously, this ideal was not realized, in part because some sites were so expensive to clean that no one party could be identified that had the requisite resources, but, more significantly, because the unnamed PRPs had a proprietary interest in the quality of the mitigation. The care taken in the cleanup could affect all PRPs' liability exposures for damage to human health of the environment. Thus, if they were to be potentially responsible for some of the costs of mitigation, they wanted to be able to participate in determining the cleanup efforts to be undertaken. As a result, the PRP named by the EPA would not be willing to initiate a voluntary cleanup until it had identified and obtained the participation and approval of the other PRPs it pursued to pay some portion of the costs. The resulting delays and associated problems of CERCLA implementation are obvious in our Chapter 4 narrative.

The US decision to avoid the problems of fault-based liability have also had the effect of increasing the risks faced by prospective new purchasers of previously contaminated sites. Under a fault-based liability approach, such as is present in the EU member-states, the owner of a previously or currently contaminated site can accept all liability for the pollution and assure a would-be purchaser that any future costs associated with the contamination, whatever their basis, will not be a charge against the new owner. The offer of such private indemnification is not possible under a strict joint and several liability regime, since the state may approach any of the individuals in the chain of title and use of a site for costs arising. Obviously, in a private enterprise system, the presence of the risk exposure created by the US approach to liability depresses the expected return on investment and retards redevelopment relative to what may be attainable under fault-based liability. This unintended negative effect on development is one of the reasons a growing number of US states are promulgating covenants-not-to-sue and providing indemnification of new landowners who are not logically responsible for past pollution on sites they are only now acquiring.

Turning now to prospective liability, we find an approach under CERCLA that could not be better designed to inhibit redevelopment and reclamation of potentially contaminated brownfield sites. Existing technological capacities and scientific knowledge are not complete. Advances can be expected in detection capabilities (which have already progressed tremendously under the impetus of the Superfund law), in environmental epidemiology, and in our understanding of the impact of different contaminants on ecosystem functioning. It is inevitable that some contamination mitigations conducted today under the best scientific evidence of needed cleanup and employing the most technologically sophisticated means now available will, as scientific knowledge advances, be found in retrospect to have been inadequate. The unique feature of CERCLA liability (which is not present in any of the other US laws regarding toxic materials, including the landmark 1976 Resources Conservation and Recovery Act and Toxic Substances Control Act) is its provision for assignment of future liabilities that may arise due to advances in science and technology.

CERCLA legislates a mandatory 're-opener clause' on any regulated and approved cleanups. In other words, no mitigation can ever be considered completed under the law. Should new knowledge indicate that the actions taken under EPA direction or conducted by the Agency itself were incomplete, all the PRPs liable for the initial cleanup would again be subject to liability, both for unforeseen damage done and for the further reclamation effort required. Any party acquiring title to a site after an approved cleanup would, itself, become liable under the strict joint and several liability provisions for the new mitigation expenses. In effect, this clause in the Act exposes all current and prospective future PRPs to unlimited potential costs. Its implication for the rate of voluntary cleanup and for investment in the redevelopment of contaminated land is obviously devastating.

The individual US states and cities, faced with the immediate need to both clean up contamination that generates ongoing health or ecological damage or poses continuing risks of such effects and to promote economic development, have struggled to overcome this constraint. Their successes have been mixed, at best, but, more important, to the extent that they succeed, their actions shift the burden of responding to scientific discoveries in the future from the nation as a whole to those sub-national public sector entities that have already been burdened with the economic loss of older industries and are the most desperate to attract new development capital.

The European approach, overall, is predicated upon state acceptance of responsibility for response to newly uncovered risks as they arise in the future. Consistent with its fault-based liability approach, which may be a much purer polluter pays principle than the US strict liability scheme, the EU treats future liabilities as state responsibilities, except in so far as past polluters can be identified as having caused the problems. New owners, except in those instances in which they voluntarily accept responsibility for mitigation of past pollution (or, in the Danish case, when they clearly buy land at below market price, knowing that the depressed value reflects existing contamination), have no prospective liability for the dangers uncovered by advancing science. Closure on costs is possible, and arises when a current cleanup is completed.

The uncertainty facing investors in previously contaminated land is thus limited by the state acceptance of what may be considered a collective responsibility to provide the necessary level of protection for citizens and the environment at large. We might be inclined to associate this acceptance of responsibility with the more corporatist orientation we have already ascribed to the EU member-states. However, the willingness of many of the individual states in the US to accept a comparable responsibility suggests that what is at work here is pragmatism, a recognition that private economic actors simply will not accept the risks that the public sector may have to accept for the welfare of the entire population.

ACCESS TO CAPITAL AND TO LIABILITY INSURANCE

The relationship between access to capital and the effects that cleanup approaches to contaminated lands have on the rate of reclamation of brownfields is fraught with irony and contradiction. Generally, it is accurate to say that the tighter the supply of development capital, the more difficult it will be to reclaim previously contaminated sites. This straightforward proposition, rather than describe a causal chain involving contamination mitigation, severs the causal link!

To the extent that funds for redevelopment are lacking for reasons other than contamination (middle-class preference for suburbs, factory single-storey building requirements, improved telecommunications spanning space, capital flight to other countries, or the like), then reclamation and reuse may never occur, regardless of the approach to contaminated land. Certainly, the policy may act as a constraint on rates and levels of reclamation effort. The critical policy question, derived from the logic of constrained optimization, is whether or not it is a *tight* constraint, the barrier that is operational and actually shapes the investment pattern on brownfield sites. Notwithstanding its strict liability policies, therefore, the US, which more readily accepts urban abandonment than does Europe, may experience *fewer* impediments to redevelopment due exclusively to its contaminated land approach than might the member-states of the EU.

For all the disincentive effects we have already identified in the US liability provisions, evidence that other contra-urbanist tendencies are not present, we cannot conclude that liability is driving capital away from urban brownfields. The liability issue discussed repeatedly in the congressional hearings may be little more than a smokescreen for avoiding admitting the true reasons why many areas of US cities are capital starved — possibly including issues of race (cf. Simmons 1991). A representative of the American Banking Association, Mr Simmons, in attributing the existence of 'brownlining' practices on government policies, may be intentionally overlooking the many other forces that are diverting capital from urban centres.

Whatever the 'real' link between access to capital and contamination cleanup, however, we have already, in Chapter 2, traced the relationship between some institutional dimensions and experiences and the availability of capital for reclamation and redevelopment, other factors being constant. We re-examine these factors in Table 7.4, noting some differences between the EU member-states and the situation in the US.

Lending practices and the processes for approval of commercial loans exhibit different patterns in the US and in the EU member-states. Some of this difference may be attributable to the smaller size of European nations, some to the historical traditions of lending institutions. The very concept of a merchant bank, for example, remains alien to most US practice. Both the EU and US are experiencing increasing bank centralization, with the US only recently permitting bank-holding companies to own banks in more than one state and both encountering the development of multinational bank ownership only in the past decade.

We need to distinguish here between formal and institutional centralization in order to understand the US pattern. Municipalities as small as 25,000 in the US may have their 'own' local bank, fully accredited and legally empowered to make investment decisions independently under government regulation. There is thus a much greater potential for genuinely local investment decision-making than is present in the largely national banking systems that prevail in Europe (with some exceptions such as the independent Yorkshire Bank in the UK). While savings institutions and building societies in Europe are less centralized than the banks, they do not routinely finance large-scale industrial and commercial redevelopment. The US, in comparison, appears to exhibit relatively low formal legal centralization.

However, the investment practices in the US reflect an institutional pattern of centralized capital allocation. The small municipal and regional banks use what are known as 'correspondent banks', large commercial banks in the major US money centres, to invest much of their capital. A common practice is the transfer of all excess deposit balances (those not fully charged against current investments under the US loan to deposit ratio requirements) *daily* to such correspondent institutions for investment. Large-scale projects (which in a smaller municipality may be any involving high risks, such as brownfield investments) are typically 'brokered' — shared out across a number of different lenders who pool their risk exposure and capital; brokering is a function of the correspondent banks. These interbank links indicate how the large, centralized institutions, although in the minority, have for decades directed the investment decision-making and spatial capital allocations of the formally decentralized banking system in the US.

A further push towards centralized lending decisions in the US comes directly from the experience of brownfield redevelopment under CERCLA. The legal complications and financial risks associated with any involvement in a contaminated property are so well recognized now that locally approved investment decisions are increasingly reviewed by specialist environmental attorneys seeking to limit financial institutions' exposures to liability risks. These specialized offices are generally at the banks' headquarters; their effect is to limit the discretion of local banks (even those with a score of branches in an urban area of one million or more) that are part of nation-wide bank-holding companies. The resulting decisions on investments increasingly ignore the local knowledge of real estate market conditions and opportunities that are the key information asset of the local bankers.

Centralization of decisions and their isolation from the knowledge and insights of local lending institutions have different effects in the US and the nations of Europe. With its fifty states and vast geographic area, the US exhibits a variety of local conditions (real estate market, public sector regulation, public participation in private redevelopment efforts, and so on) that simply do not exist within the more limited geographic areas of the individual EU member-states. Investment decisions that ignore or supersede such local considerations with risk-avoidance-driven standardized and centralized policies are less likely to be optimal. The Superfund, therefore, may be charged with having generated institutional behaviours that undermine the productive and profitability of lending practices in the United States, in addition to making brownfields reclamation more difficult.

The somewhat greater acceptance of risk in general in the US relative to Europe provides a counterbalance to the effects of this central screening of brownfield loans. Venture capital itself may flow to high-risk, high-potential return urban regeneration projects that are intended to be liquidated once redevelopment is completed. A further countervailing factor is the US public practice of providing loan guarantees for certain types of development projects. (The federal and state banking regulations that require deposit insurance provide protection to those placing funds in the institutions, not the banks themselves.)

The extent to which general liability insurance covers the unanticipated retroactive liabilities of landowners and users for past contamination affects the capacity of those parties to conduct their own mitigations of their properties, whether or not they seek additional funds from lending institutions. As we have seen, litigation over the extent of coverage intended under business general liability policies has emerged under the provisions of CERCLA in the US. While famous for its willingness to insure against high risks, Lloyd's of London has suffered significant and substantial losses, some involving contamination; the availability of private sector coverage for high-risk exposures has been shrinking over time on both sides of the Atlantic. Any reduction in the ability to hedge risk with purchase of insurance will, in the absence of other changes, tend to depress investment in uncertain projects, and thus reduce the rate at which private-sector-led voluntary contaminated land cleanups occur.

Finally, we need to consider the impact of the actual experience of capital flight on redevelopment efforts, and especially on the willingness of the 'local' state (a US state or EU member-state) to take steps to promote brownfields reclamation and redevelopment. The differences between the US and the EU in this dimension are somewhat smaller than the similarities. Both have extensive dereliction

Table 7.4 Access to capital and to liability insurance

	'Europe'	'North America'
Private sector lending practices	Central office control over lending decisions	Substantial municipal-level branch bank decision-making
Centralization of financial institutions	Formally and institutionally high; limited local branch lending discretion	Formally low, albeit growing; institutionally high; central loan clearance growing for brownfields
Financial institution risk aversion	High, especially in banking sector	Middle; willingness to speculate
Recent experience with risky loans	Minimal losses experienced over land contamination	Major writeoffs due to efforts to avoid liability exposure
The availability of lending insurance	Low; brownfield project risks limited by state policy	State-required deposit insurance; limited public loan guarantees
General liability insurance practices and experiences	High risk-taking at Lloyd's; losses causing reversals of policy	Contamination not previously covered; contentious issue
Experience of spatial displacement and capital flight	Localized domestic displacement due to small nation size; significant international relocation	Domestically widespread; extensive inter-state locational competition
Actual losses to external locales	Significant in specific local instances	Intra-metropolitan relocation as well as inter-state
The extent of land dereliction and abandonment	Widespread in about half the member-states	Spatially concentrated in older regions

and abandonment, possibly if not probably accompanied by past land contamination, concentrated in some areas and virtually non-existent in others.

In the EU, the member-states that led the industrialization of Europe have the most acute contamination and redevelopment needs, notably the UK, Germany and Belgium. Germany, in particular, is faced with the legacy of under-capitalized and environmentally negligent production in what was the German Democratic Republic before reunification. However, some of the most affluent member-states have little problem with domestic dereliction, notably the Scandinavian countries. Thus there is only a limited correlation between the incidence of land contamination and population affluence.

The US has similar concentrations in particular states, predominantly in the industrial belt of the Northeast and Midwest, accompanied by California. By contrast, however, the states with the largest concentrations of major land contamination (reflected in the number of federal Superfund sites) also tend to have per capita incomes that are above the national average. A national policy that leaves the states the option of accepting some risk of liability as a means of promoting reclamation and reuse of brownfields may be eminently appropriate: the populations that have benefited the most on average from past economic activity that produced pollution have, as a means of continuing their economic development, accepted potentially higher costs for cleanups than will be charged (by the national state alone) to the population as a whole.

Older industrial and commercial areas in both the EU and the US have also suffered capital flight and abandonment. Shortages of private investment funds increase the intensity of competition for those resources, so the relative inferiority of polluted sites and their limited ability to attract development capital heightens the need to systematically channel public funds to their mitigation and redevelopment. The US, however, faces a further problem, derived in large part from its multiplicity of small political jurisdictions and decentralization of political power: intra-metropolitan capital flight.

The American reclamation dilemma is not one of attracting capital to depressed older conurbations, but, rather, to retain or rechannel capital into central city locations that have been abandoned in pursuit of more intensive development of suburbs that are within easy reach of the metropolitan centre. This additional level of competition for capital intensifies the problems faced in attracting private sector capital to inner city brownfields. While major European conurbations may face a similar problem spatially if greenbelts do not limit the expansion of the city, it does not take the form of competition between political jurisdictions for capital. Land-use controls and zoning powers can be coordinated across an entire urbanized region to encourage investment in brownfield sites by limiting access to or uses for alternative development properties. With its inter-municipal competition (and extremely heavy reliance on real property taxation to fund the local state), the US is incapable of similarly channelling developers towards contaminated lands.

Overall, it appears that the institutional settings in which property and development investment alternatives are generated and then assessed by the financial community are more problematic in the United States than in Europe. Whatever the provisions of US law and regulations that control contaminated lands and their mitigation, their capacity to contribute to urban regeneration is impeded by these problems in capital allocation and access.

CURRENT POLICY AND POLITICAL PRESSURES

Turning to the future, and considering how the experiences, conditions and institutions we have thus far discussed will shape emerging policy towards contaminated land, we need to take current political factors more directly into consideration. The obvious factors here are, first, recent experience under whatever laws and regulations are in effect, and, second, the strength of those political actors that favour strong environmental concerns. (We presume that the forces promoting economic development and those representing private industrial and commercial interests remain strong and influence the political process in any event.)

Experience under Current and Past Cleanup Efforts

In the United States the delays experienced with the Superfund site cleanups, the massive litigation that has resulted among the potentially responsible parties, and other aspects of contentious cost-sharing have combined to generate a poor image for CERCLA. The much larger numbers of contaminated sites that have been cleaned up are often overlooked, especially in the congressional hearings, as Chapter 4 indicated. Similarly, the high publicity over the putative effects of CERCLA liabilities on urban regeneration has overshadowed the many successful inner city reclamation projects that have proceeded, some even on clearly contaminated lands. Interstate competition in the US for industrial relocations and new jobs may have generated tendencies towards the states undermining standards, but the federal statute and regulations leave them little room, other than to subsidize progress on cleanups. This sub-national response may leave the more heavily polluted states bearing costs for actions from which the entire country benefited, but it also means the states with above-average personal incomes are the most heavily burdened. Furthermore, concern for environmental 'justice' in exposure to environmental health risks has focused attention on race and income equity considerations in the siting of new hazards and in the rate of mitigation of the old. Since this issue has led to *greater* federal concern for mitigation in impoverished, but neglected, areas, urban regeneration has become a more important concern at the national level as well.

The US standards, while strict, are in many dimensions weaker than those in Canada, the nation's largest trading partner. Yet Canada has learned to clean and avoid contamination without the conflicts prevailing in the US. The lesson from its northern neighbour may shape future implementations of contaminated land policies in the US. The southern neighbour, Mexico, however, has far weaker standards, and tends not to enforce those standards it has. It is a measure of the political will and potential behind the cleanup movement in the US that, when the extension of the free trade pact with Canada to include Mexico in the North American Free Trade Act was proposed, a side agreement on environmental protection, in addition to one on workers and fair labour practices, was necessary to obtain congressional approval for the pact.

The European Union treaty provisions now *require* that environmental policy considerations be incorporated into *all* EU policy. Thus policies for generating higher environmental standards in general, and contamination avoidance and mitigation in particular, are beginning to 'bite' and have a real impact. All EU-funded projects under its structural funds, for example, have to meet environmental standards. Since these capital projects are often in now relatively derelict or underused former industrial and mining areas, this requirement will have the effect of forcing more rapid mitigation of past contamination as a condition for economic regeneration if EU funds are employed.

Some have argued that the bulk of new member-state environmental legislation is strictly EU-driven, given the many environmental process and standards obligations embedded in the Fifth Environmental Action

Programme, *Towards Sustainability* (CEC 1992a). Industry in Europe, is getting a similar message: that public opinion supports stronger environmental standards and that objecting or resisting environmental regulations is bad publicity, although this message may be better received in some countries (e.g. Germany) than in others (notably, the UK).

The Strength of Environmental Movements and Green Parties

Europe has 'green' parties; the US does not, at least at the national level, although local environmental issues have proved to be a useful political platform for some individual candidates. The Greens in the UK are mostly a lobby, and do not act systematically as a political party; even so, in the 1989 voting for members of the European Parliament, the Green vote was 15 per cent. An avowedly Green party presence asserts itself in the European Parliament, albeit with only some five per cent of the seats, and Greens have held seats in the national legislatures of France, Belgium and Germany, often participating in ruling parliamentary coalitions. Thus the formal political base for environmentalist politics is unquestionably stronger in the EU than the US.

Environmental lobbies have been strong in the US, and the growing environmental consciousness and accelerated passage of major environmental legislation discussed in Chapter 3 suggests the politicians have been listening. However, the continued support for national-level efforts may be a function of the rate of recurrence of highly publicized environmental problems or adverse impacts, including those at the local level. Anti-regulatory industry interest groups have responded by financing and nurturing their own local interest groups, attracting workers to their cause by highlighting the potential threats to their employment posed by environmental controls. In political and legal orientation, these groups challenge the environmentalist premise that the uncontrolled pursuit of economic gain produces socially negative outcomes (Jacobs 1993, 1994). Their growth has been linked to the broader right-wing shift in US politics; major actors in the 'Wise Use' and 'county movement', as these groups have labelled themselves, have even been linked to overt acts of (arguably terrorist) political violence (Helvarg 1994).

The Republican-controlled Congress that took office in January 1995, may be expected to be more responsive to these new anti-regulationist groups. It is difficult to project probable directions. On the one hand, one Washington policy analyst observed, just as the new Congress sat in its first session, that the only agency not mentioned as a target of cuts (other than the Defense Department) was the Environmental Protection Agency. The Republicans' 1994 'Contract with America' political platform was designed to reflect public opinion; the exclusion of the EPA from the visible cuts list reflects the strong remaining public support for environmental causes and, especially, local contamination mitigations. However, there is an enduring reported association between political liberalism and environmental concern in the US (Mohai and Twight 1987; Scott and Willits 1994; Van Liere and Dunlap 1980). Thus growth in the growing conservative trend in US politics may presage a reversal on contamination policies.

The activation of the European Environmental Agency in Copenhagen after some delay will certainly affect local environmentalism throughout the EU. The Agency provides a focus and direction for supra-national appeals by locals of planning decisions they consider inappropriate. Since opposition to roads and other local infrastructure building is often based on a wide variety of grounds, the birth of the EEA could increase reliance on environmental rationales and thus raise environmental consciousness. Another factor affecting the EU is the imposition of common standards under the Single European Market, which increases the pressures for environmentally responsible behaviour by businesses intending to appeal to the whole continental market, since the EU expanded in 1995 to include more environmentally conscious member-states.

Overall, we suspect that the prospects for continuation, if not strengthening of high standards for environmental performance by both the public and private sectors, are stronger in the EU than in the US. However, it is difficult to assign this probability to particular institutional conditions or forces other than the difference in the political visibility and thus power to put issues before the public associated with explicitly multi-party parliamentary rather than implicitly two-party presidential political systems. Discussion of such systemic differences is, however, beyond the realm of our review here — and the 1992 experience with a third party gaining almost 20 per cent of the votes in a US Presidential election suggests that we cannot rely too much on this distinction!

THE COMPETITIVE POSITION OF CONTAMINATED LANDS

Our reviews of the policies of the US, the EU and a number of EU member-states should have made it very clear that the presence or threat of contamination creates problems for would-be developers, especially in urban areas. However, the overall planning context and especially the narrowness of zoning and rigidity of the land-use controls prevailing in a given legal setting have a large influence over the effective costs of new land-use regulations, such as those for the mitigation of contaminated or potentially polluted and damaged properties. Our point is simple: while the possibility of land contamination necessitates that certain measures be taken in assessing development prospects and the presence of contamination may lower investment returns, the significance of the different costs may vary. If the existing land-use control system is sufficiently strict, those costs may be incremental, relative to total project size. If there are virtually no land-use controls and the presence of contamination puts such constraints in place, they may have a very significant effect on total project appraisal and profit potential.

Recalling the cost elements we introduced in Chapter 1, we compare the US and the EU overall in Table 7.5. Where necessary, we comment on variation among the EU member-states in Column 2. Overall, however, we have a rather obvious finding, deriving from the differences in the patterns of land-use controls in the EU and the US: the potential financial disadvantage of contaminated sites is far greater in the US than in the countries of the EU. Furthermore, the liability scheme prevailing in the US heightens risk and further reduces the competitive appeal of brownfield sites.

Looking at the transaction costs, we note that, whatever the costs for environmental assessments of properties prior to their sale (or prior to redevelopment), they are less likely to cause problems for developers by delaying projects in the EU member-states than in the US. Far more rigorous planning standards tend to occupy more time even in the absence of site assessments. Thus, if assessments are conducted, they may be undertaken simultaneously with other planning actions. (Of course, if developers and their financiers fear total rejection of their plans, they may defer the site studies until after they get planning approvals, in which case, they could face the same delays as are common in the US.) The legacy of strict planning in the European cases, however, may reduce the repetition of underwriting by different

Table 7.5 Investors' costs of contaminated land policies

Cost factor	European Union	United States
Transaction costs		
Environment assessment fees required for sales	Currently varies with national liability rules	High, unavoidable for financed sales. and could reach $10,000/0.25 acre
Project delays	Long planning process leaves time for site land assessments	Speeding planning reviews slowed by new site assessments
Increased project underwriting; allowance for unpredictable cleanup costs	May occur; risk reduced by centralized lending standards and high quality records on recent land uses	Significant due to emerging standards; lender standards vary with their fears of risk of liability exposures
Costs reducing perceived returns on investments		
Uncertain mitigation costs	High; liability assignment is common; state may pay all	High; mitigation standards rise over time, raising risks
Liabilities for damage done by past pollution	May occur; court cases such as Cambridge Water in UK may limit risk for past actions even if responsible	Potentially unlimited regardless of fault; lender liability risks are still not fully clarified
Property stigmatization	Possible; urban land values may make it irrelevant	Perceived to be high, with minimal evidence
Developers' reputational losses	Important in UK; not so in most of Europe	Totally irrelevant in US capital markets
Future deed restrictions and monitoring requirements	Alternate land uses are already restricted; monitoring may cause some constraints	Both may limit future land uses due to action by new owners or any level of government

Future prospective liability if mitigation or health risk standards change	Irrelevant to private developers; state responsibility	Major cost uncertainty to developers; high risk of future cash flow drain
Risk of effective asset seizure for costs and charges due the state	Expropriation is a risk, but compensation is guaranteed for innocent lenders	Significant risk factor for private lenders if governments are involved

Note: These findings are for non-NPL sites, the vast majority of all instances of development of previously used and potentially contaminated lands.

financial groups and add certainty to cost projections since those controls generally make better records of past land uses available to decision-makers.

Even greater differences are evident in the costs affecting — that is, lowering — the expected rates of return on redevelopment investments. First, developers in both the US and EU face potentially very high mitigation costs in their brownfield projects. In the EU member-states, however, provisions exist for uncertain future liabilities to be accepted by buyers or for them to be accepted by the state. CERCLA's joint and several liability provisions leave any developer liable for future costs for which the national state does not accept responsibility; competitive economic development pressures have, however, led some individual US states to accept those liability risks.

The Superfund also places no limit on potential contributions to pay for past damage done in the US, regardless of blame, under strict liability. The UK, exemplified by the Cambridge Water case, places no blame for damage done without intent, even if blame can be assigned. The Netherlands' practice of having the state prepare land for development shifts such burdens to the state. In general, the EU does not now allocate liabilities for such past acts on current owners and developers. The absence of a formal EU standard for contaminated land policy, however, could change this pattern in the future.

Property stigmatization is, in effect, a luxury. The smaller, denser, states in the EU cannot afford the luxury of stigmatizing and abandoning land, especially that served with well-developed infrastructure, to the extent that is possible in the US. This implies, however, that brownfield sites that may carry any of a number of additional costs when compared to greenfield locations for development will always be more difficult to regenerate in the US than in the EU, regardless of the details of the contaminated land policy. Thus, the EU need not fear the rates of abandonment and effective 'brownlining' of brownfield areas, which have become evident under CERCLA, even if the EEA and environmentalist member-states lead the union to consider similar legislation.

The peculiar importance of reputation in attracting business capital in the UK may have an effect on risky regenerations not felt elsewhere in the EU. It may explain in part the UK reluctance to accept the Lugano Convention and other toxics policies put forward by supranational bodies. On this level, at least, the US does not face regulatory impacts on its property market processes.

The tight control on alternative land uses that is commonplace in Europe is virtually unheard of in the US. Thus any outcomes of the discovery and treatment of contamination which lead to a reduction in the range of possible future uses of a site will have automatic and substantial impacts on project valuation in the US, but analogous findings may have no effect whatsoever on the range of already constricted alternative uses in many European countries. As a result, the expected reduction in project valuation associated with suspicion of some prior contamination will be far higher in the US than in the EU states.

Prospective damage or cleanup liabilities pose major uncertainties. The distinction between the collectivist and individualist orientations of the EU and the US present would-be developers with very different risk conditions on this factor. The member-states of the EU by and large all seem to accept state responsibility for costs resulting from the discovery of new knowledge that leads to reassessment of health risk or the adequacy of prior mitigation actions. The US, although it is a leader in generating that new knowledge, appears to persist in placing the burden for conforming to new, generally higher, standards on the individual chains of potentially responsible parties for each individual site that is affected by a new standard.

Individualism produces a counterintuitive result in the final cost element, the likelihood of an investor losing an asset or collateral due to state action. In an economy that lionizes individual institution risk-taking, lenders may lose their collateral in the US for taxes and fees due to the state, which is an extraordinarily high risk given the treble damage penalties possible under CERCLA. By contrast, the collectivist ethos still prevails in practice in most European countries, notwithstanding their rhetoric; it remains a factor even in UK policy-making despite a fifteen-year privatization drive. Such corporatism leads to state protections for lenders when public actions are taken against their borrowers, under the principle that no one innocent party should bear costs for actions taken by the state on behalf of the public good.

In countries with either an acute shortage of land or a relatively even distribution of land contamination, the factors raising costs for those who would redevelop brownfield sites are less important than in settings with abundant real estate investment alternatives. However, capital flows do not stop at national boundaries. Thus the Single European Market has created a situation within the EU that is analogous to the inequality of historical pollution in the US. Land development capital may, if regulations and financial risks get high enough, flow from what we have called Group 1 towards Group 2 countries within the EU, from the states that led the industrial revolution to those that did not develop their heavy industry as intensely in the period before we understood about pollution as well as we now do. Similarly, as the exchange rates between Canada, the US and Mexico stabilize (and as NAFTA is extended south), the competition already evident internally to the US may extend across national boundaries.

The emerging issues then move from maintenance of high standards — which would permanently impair the redevelopment prospects of virtually all high-intensity prior use brownfields — to one of the cost allocation of the burden of such standards. International standards alone may not be sufficient. For them to be enforced — and for private capital to bear some of the costs of redevelopment altogether — supernational funding for some share of the costs of cleanup may be warranted. This is a principle that is the industrial nations' internal parallel to the demand from industrializing states that those already developed pay them for their sacrifices in retaining rain forests and other global common pool resources. If the cities are such common pool resources themselves, then it is appropriate that such a principle apply.

8. Lessons for Future Contaminated Land Policy: Prospects and Pitfalls

Considering the experiences of the EU member-states and the US, it appears obvious that ecological integrity, urban well-being, economic regeneration and public health, all objectives of efforts to clean up contaminated land, cannot be accomplished by the public sector alone. In fact, as Healey and Nabarro (1990) note, 'investment interests, the providers of finance for urban development and property purchase, increasingly drive the process of urban change'. Their comment reflects the growing power of the private sector in late twentieth-century capitalism as individual countries, not to mention sub-national governments, are powerless in the face of increasingly international capital markets.

Whether this growth in private power is desirable or not is beyond the scope of our examination of contaminated land issues. It is part of the economic reality within which public policies must be formulated. We thus turn here to derive some findings from our comparative assessment and suggest directions for public policies that can protect human health and the environment at large from land contamination while simultaneously attracting private capital to reclamation efforts.

We begin by summarizing the experience to date, noting that the frame of reference we may adopt, or the burden of proof we impose, leads us to disparate conclusions on 'success'. Next, we frame our criterion for judgement: the impact of contaminated land policy on the reclamation and sustainability of the common pool resource we have called the urban milieu, recalling our discussion in Chapter 1. Using the concept of the CPR, we then posit some principles for managing the urban resource, identifying what we consider to be elements of an appropriate cleanup policy. Next, we address the risks associated with the presence of land contamination and with policies to address it, isolating elements of risk, different parties exposed to various mixes of risks, and addressing the processes of risk perception and risk aversion.

These brief overviews are essential to our discussion of current pressures for policy change and our concluding section on the probable future of contaminated land policies in the industrial nations that have fouled their own nests in their past pursuit of affluence.

THE EXPERIENCE TO DATE: A SUMMARY ASSESSMENT

A quick summary of experience on both sides of the Atlantic is exceedingly difficult. Mayor Koch of New York City was famous for his question to all who would answer, 'How am I doin'?'. Any answers to the analogous question about public policy towards contaminated land — how are they doing? — vary with the perspective and data sources of the respondent.

If we were to draw solely from news media accounts and public policy debates, we would be inclined to declare existing policies to be, by and large, failures. Horror stories abound about discoveries of contamination, unexplained illnesses shared by residents of, or workers at, particular sites, and high potential economic development projects that died on the shoals of contamination risk. All are reported regularly in the media.

If we looked at the policy development process and the levels of state effort that have been committed to identifying the problem of contamination and responding to it, we would have to record significant progress. The US has passed extensive legislation and is committing substantial financial resources to cleanup; sites *are* being cleaned up, some quite rapidly. The EPA, hobbled in the first eight years of CERCLA by the Reagan anti-regulation team, has begun to make real progress, and the states themselves are confronting both the environmental and economic development problems directly with their own laws and approaches. The EU, while it has a much shorter history of addressing contaminated land, is accelerating its efforts. The UK opposition to much EU policy has isolated the country among its fellow member-states and the recent enlargement of the EU holds the promise that land contamination will soon be defined as a genuine European issue. The issue of brownfield — and more generally, urban — redevelopment figures prominently in the EU's *Fifth Action Programme on the Environment* (CEC 1992a).

If we address the question from the perspective of the actual progress on the ground, we have substantial successes to recount. The examples in Chapter 4 and 6 suggest real progress is being made. While the US policy debate bemoans the slow pace of Superfund cleanups, those roughly 1,300 sites are a small fraction of one per cent of the total polluted sites, and many of the hundreds of thousands of smaller and less serious contamination cases are being addressed, largely with private capital. Europe's successes involve a number of very large projects, spanning whole contaminated regions.

Finally, if we ask whether contamination is actually being reduced in aggregate, we confront a conundrum: new sites are being discovered more rapidly than the old are being reclaimed. 'Contamination' is, in effect, a moving target. The passage of US hazardous waste legislation, dating to the mid-seventies, has spawned a detection and mitigation industry that, throughout the industrialized countries, continues to advance the technological capacity to both identify toxics in the soil and render them, if not completely safe, far less dangerous. The National Research Council's 1991 finding that 40 million Americans live within 4 miles of a Superfund site could not have been made a decade earlier — the data were not available. Depending on the rates at which Superfund sites are identified and mitigated, this number could rise or fall; either result could represent progress. The attention focused on contaminated lands has heightened community awareness and the political will to respond to threats to human health and ecosystem functioning. The sometimes very narrow and spatially confined impacts of land contamination have contributed to a recognition of a new form of inequality: exposure to environmental risks.

We thus find it difficult to answer the question of how different countries have done with their contaminated land policies. We are confident in asserting that, on balance, the rate of new contamination per unit of output has been reduced, and the extent of human and ecosystem exposure to past pollution is being eased. Certainly in the US, the combined financial risks embedded in TSCA, RCRA and CERCLA have successfully placed hazardous waste reduction or prevention ahead of disposal on the agenda of American business. If for this reason alone, the draconian measures in the US legislation will make the world a cleaner place than it would otherwise have been. Whatever their faults, contaminated land policies have affected business behaviours. As an executive of a major US paper products company has observed, 'We can no longer rely on capital outlays and solutions to deal with waste after it's produced. The new frontier is waste prevention' (Larson 1992:31).

No reports of policy success, however great, can be presumed to constitute a finding that existing approaches are as strong, effective and efficient as they may be. The examples of apparent successes and failures on both sides of the Atlantic suggest a variety of principles that could serve to guide improved contamination land policy in both the US and the EU.

PRINCIPLES FOR MANAGING THE URBAN COMMON POOL RESOURCE

If we consider contaminated land policies to be efforts at improved management of common pool resources (CPRs), we can then follow Singh, once again, as we did in Chapter 1, notwithstanding his analytical focus on CPRs in Third World natural environments. (In so doing, we follow a pattern evident in other efforts at environmental conservation, in which the industrialized world continues to learn from peasant societies.) Specifically, CPR management (CPRM) constitutes 'human intervention in a CPR system with a view to restoring it, or conserving it, or augmenting and sustaining its productivity, and/or regulating its use' (Singh 1994:7).

Admittedly, the details of Singh's management discussion are not directly transferable from rural natural resources to the resources inherent in an urban area. However, his principles still apply. The four management objectives he posits can be recognized as those of the policies we have described: survival of human and other life systems, sustainability and efficiency, equity, and pursuit of national economic self-interest. Even the three major approaches to CPRM policy he compares, privatization, centralized public management (nationalization) and decentralized collective management (community control), comprise alternatives for contamination management in modern urban areas that we have observed, albeit not in exactly the same terms.

Privatization is evident in US policy that attempts to assign individual responsibility for past contamination, places trust in private entrepreneurs to provide the cleanup required, and has, despite legal requirements to the contrary, tended to give relatively low weight to community concerns. However, the city as a whole cannot be privatized. It is, by definition, an aggregate of large numbers of individuals and firms that share a common location in space and the environment they collectively create at that location. Arguably, then,

CPRM applied to urbanized areas is inconsistent with privatization of responsibility for contamination that affects the value of the whole CPR. The urban revitalization problem, the issue that drives the desire to redevelop, not just clean up, contaminated lands, is an urban CPR issue. It is not surprising, then, that a privatized approach to cleanup generates conflicts over cleanliness standards, selection of mitigation techniques, and responsibility for assuring effective implementation of reclamation plans. Such conflicts have repeatedly been documented in US Government Accounting Office reviews of CERCLA and its implementation (US GAO 1992, 1993, 1994e).

Nationalization of either contaminated sites or of responsibility for managing the urban CPR is rare. Such practices as The Netherlands' reliance on the national state to prepare all sites for future use — and to preclude private economic gain from real estate development — come close to a nationalization of the urban CPR. The smaller member-states of the EU may have little alternative to nationalization, since they have, in effect, virtually a national CPR in their overall environment. Within the larger member-states, acceptance of state responsibility for mitigations is evident in the German approach to the Ruhrgebiet or Bitterfeld, Italy's designation of UTRAS (Territorial Units of Historical–Cultural and Environmental Recovery), while land acquisition and reclamation of urban contamination has been one of the objectives of the UK 'Garden Festival' programme (Meyer and Burayidi 1992). This evident concern for approaching the contamination problem as a CPR management task is consistent with the corporatist orientation which we have already noted characterizes the European approaches.

Ironically, we look to the federal takeover of the management of sites that is characteristic of transfers of contaminated lands onto the National Priority List in the US Superfund programme for a form of selective nationalization of mitigation management. At a minimum, it is acceptance of centralized public management responsibility for selected sites that either impose massive physical threats to the local CPR or cannot be managed otherwise. Such a centralized approach also would, much like privatized cleanups, tend to ignore local concerns and considerations. Superfund thus has been attacked simultaneously from the left for ignoring the people 'on the ground' in communities suffering contamination's effects on their CPR, and from the right for failing to address the concerns and well-being of local business interests and the individual potentially responsible parties (US GAO 1994b, 1994d).

Both nationalization and privatization of contamination cleanup efforts carry with them the high risk that community interests, that is a concern for the urban CPR, will not carry much weight in decision-making. Under either approach, pursuit of 'efficiency' in pollution mitigation, in bringing derelict land into the market, or in attracting private parties to reclaim and redevelop contaminated lands may be expected to give only minimal attention to the impact of policy choices on the urban CPR. This is to be expected.

The objective of a nationalized approach is the greatest gain for the nation as a whole; to the extent that the condition of land in any one urban area is not essential to national well-being, it is not worth the expenditure of special resources. The sites which end up on the NPL in the US, for example, are only those that both threaten health or the ecosystem with risks that either are so large they cannot be ignored or have the potential to spread to other areas, *and* have potentially responsible parties who cannot agree on actions to take or how they will share the cost burdens. While mitigation of the problems at such sites certainly affects the local urban CPR, the primary concern is the non-local effects of continued contamination.

Fully privatized mitigation efforts are governed by a profit motive, even if the cleanup is motivated solely by the threat of public imposition of damage claims or other costs. Thus private cleanup efforts operate on the basis of cost-minimization, and damage to the local urban CPR is relevant to that accounting logic only in so far as that harm may impose costs on the private mitigator. Since the local state cannot sue over a decline in its CPR, the effects of the selected approaches on amenity values and other components of the CPR enter the process of selection of reclamation approaches only when the private mitigator itself partakes of the local common pool resource. To the extent that the pollution to be abated is the result of past activity, many private parties undertaking to clean up the residues of their past polluting activities may no longer be in business in the locales they contaminated. They thus have no reason to include the effects on the local CPR in their decision-making.

Decentralized collective management is thus, arguably, the preferred approach to not only management of the local CPR, but also to the contamination cleanup effort itself. On one level, it is clearly evident in many of the approaches to contaminated lands we have addressed. The management is clearly collective in the US in cases of multiple users of contaminated sites: any group facing joint and several liability which acts together to minimize their costs and the negative spillovers

Lessons for Future Contaminated Land Policy: Prospects and Pitfalls 181

from a polluted site can be characterized as exhibiting this approach. However, it is here that the problem of the membership of the management collective arises. The collective for purposes of managing the CPR as a whole, the urbanized area, is different from that appropriate to managing an isolated contaminated site within the city. This form of management logically implies participation by all parties whose well-being is affected by the way in which the CPR is addressed, and thus requires broad citizen participation, or what has come in the West to be labelled 'community control'.

Contrasting contaminated land management and CPRM as applied to an urban area, then, a site-specific approach may be said to be inconsistent with a decentralized collective management approach. The focus on single sites translates into a concern for the owners and operators of facilities on those sites, and leads to a lowering of concern for the spillover effects on the neighbourhood or the CPR associated with the particular land uses or the contaminants on the sites. Worse yet, from a collective management perspective, it fails to place any sites' neighbours at the table for discussion and decision-making on alternatives.

ELEMENTS OF AN EFFECTIVE CONTAMINATED LAND POLICY

We identify here some key elements of a contaminated land policy that could build and improve on those legislated or otherwise implemented programmes now in operation. We cannot offer implementation details for these broad elements, as the structural and other dimensions in which the policies have to be implemented are so varied (see Chapter 7). However, moves in these directions, in our estimation, will yield improved policies and reduce the negative impacts on quality of life associated with the inherited debts of land contamination and the cleanups that are needed to retire those obligations.

1. *The state needs to accept a primary role.* Leadership from the central jurisdiction is necessary to permit community participants in different localities to become involved in some form of decentralized collective management of their urban CPRs. This role, as the EU cases demonstrate, can both hold down expenses by centralizing and coordinating communications and actual mitigation, and can also assure that the subsidies provided do not

produce windfalls for even the innocent landowners of sites determined to require cleanup. The Clinton Administration's proposals for amending CERCLA greatly extended the provisions for broad-based community bodies in NPL decision-making and specified the breadth of participation from different community sectors required in order to assure representativeness. However, stimulating such local community participation may also involve changes in the role of the local state, since the affected community may not correspond to any local political jurisdiction. Care must be taken not to bound local impact considerations by the arbitrary lines that characterize the territories of units of the local state.

2. *The state needs to look outward as well as inward, and to improve coordination within free trade areas.* This is the issue of a level playing field to which we have alluded. While the US has not addressed this issue as directly as have the member-states of the EU, US businesses have expressed concerns about the standards they face for mitigation. The contention is that the CERCLA liability standards and costs disadvantage American businesses relative to European competitors (Portney 1992). If valid, and we believe the issue is more real than imagined, it suggests that the playing field argument needs to extend beyond specific free trade areas. The increasingly global capital markets from which local areas must draw for development capital dictate a policy that is trans-national in scope at the same time as it is very locally-sensitive in detail and actions. (Pearson, 1987, for example, found that strict environmental standards, writ large, did not necessarily reduce national economic well-being — but he did not address local community impacts, and the inequality that could be generated within a national economy.)

It therefore follows that, to avoid capital flight to other nations, an effective long-term contaminated land policy needs either international protocols or explicit waste-related tariffs (taxing firms producing outside the nation or its free trade area — level playing field — in proportion to their toxics generation in other countries). Toxics disposal in Third World countries will not only stimulate capital flight and threaten jobs in industrial nations, but continue the practice of using the ecosystem as a global waste sink, with possible consequences for the global CPR, the ecosystem as a whole. Both the US and the EU have policies supposed to reduce international transfer involving toxic wastes, but their effectiveness is questionable (Hofrichter (ed.) 1993).

3. *To the extent that private investment in urban regeneration is considered to be a social or public good, contaminated land policy has to increase investor certainty, or at least reduce some risk and cost concerns.* As we noted in our discussion of parties exposed to risk, those entities that are the most central to private financing of reclamation and reuse accept risk voluntarily. These risk takers, however, have the power to select which risks to incur, and it is in this context that excessive uncertainty or exceptionally high downside loss potentials can deter private investment. In the absence of publicly-led cleanups, the failure to attract private capital to brownfield sites will force continued involuntary risk exposures for the residents and workers in neighbourhoods abutting contaminated lands. One obvious mechanism for reducing uncertainty and downside risks is the conduct of state-led cleanups, preferably led by a level of the state that is responsive to local community concerns and fears. However, the US experience counsels that exceptional attention must be paid to assure that the private firms given contracts to prepare and execute the cleanups do not overcharge and inflate cleanup costs. A minimalist state apparatus that can only contract out for cleanups and not conduct them itself has to have powers to address the fact that the state and the private sector have diametrically opposed interests: contractors want to expand cleanups and increase their revenues; the state, at any given level of mitigation target, wants to minimize costs.

4. *'Polluter pays' must be considered the appropriate governing principle for cleanups.* However, this dictum must encompass the consumer who elects to use a product generating toxics in its production. Thus, if current disposal requirements raise prices for current consumption and that is 'fair', then the consumers of goods in the past should also pay some of the accumulated contamination debt. Such past consumers are, presumably, in large numbers deceased, as may be the investors who made extra profits off activities that contaminated the land (whether or not the action was taken knowingly). An equitable implementation of the polluter pays principle, therefore, should not burden only

the firms engaging in contaminating activities or those that might produce pollution. Some of the tax incidence has to be spread across other parties in the national economy. Our discussion of past consumers and producers helps to further define the appropriate incidence for contamination mitigation taxes: revenue generators with rates of taxation that rise with current wealth and with inherited wealth in particular. We have no measure of the extent to which a household or company has benefited from past pollution and thus needs to pay more under the polluter pays principle other than the wealth the unit — or its predecessors — has accumulated, given that possessions themselves involve pollution and that wealth derives from past income.

This principle also suggests that private liability for past pollution be limited, especially for acts of contamination that were acceptable at the time they were undertaken. This proposition reflects the House of Lords' finding in the Cambridge Water case in the UK. It leaves open the possibility that firms that polluted knowingly — and kept their knowledge of the damage they were causing from the government — may be excused since the ignorance of the state made their actions 'acceptable'. The EU's draft statement on liability for contamination suggests, however, that Europe may be drifting towards the stricter retrospective liability that is applied in the US under CERCLA. This may be a rational response to a tax system that is not sufficiently progressive with respect to wealth to really permit the application of a genuine polluter pays principle through the use of general (or earmarked) revenues; having a disproportionate share of the public sector cleanup cost financed by taxpaying working people and lower-income households certainly violates the principle. An inability to levy an equitable tax may have led to the political acceptability of broad impositions of costs on businesses and past and present property owners and users, with the negative effects on reclamation and redevelopment we have already discussed.

5. *Taxation and revenue-generation for financing the public sector or state share of contamination cleanups should follow the logic of the polluter pays principle.* Therefore, given the declining progressivity of most tax systems, the state share of cleanups should either be financed through targeted taxation of key polluting industries (as is the practice in the compartmentalized

'Superfund' in the US), or through increased reliance on higher progressivity in inheritance and other property transfer levies. In either instance, one of the objectives of providing for some state share should also be a reduction in the waste associated with private sector disputes of contributory shares due (as in the litigation among PRPs under CERCLA in the US). Central financing may be more efficient in terms of acquiring funds; it becomes vastly more efficient if it also eliminates these cleanup transaction costs — and it also contributes to greater investment certainty by eliminating the variable of litigation outcomes from the investment decision. On all these grounds, then, the public sector should play a lead role in conducting cleanups and mitigations in preparation for redevelopment and reuse. But this role can be played equitably only if the incidence of new taxes conforms to the polluter pays principle.

Any taxation of business or imposition of cleanup costs on private firms, if the playing field within a free trade area is level, will be passed on to current consumers. Is this any different from having *consumers* — not businesses — pay directly? Depending on the products involved, the incidence of the higher costs across consumers of different income levels can vary tremendously. If consumers are taxed directly, the state may be able to better control the incidence of the costs. (We ignore here the argument over the greater efficiency of private- or public-led cleanups since the data are generally inconclusive.)

Non-business taxation can be progressive, and thus capture some of the uneven benefits garnered by different current economic actors from past pollution. A value added tax, VAT, is reasonable if the objective is to tax current higher consumers more than lower ones, especially if the tax is not levied on certain necessities that the poor consume in greater proportion to their incomes. Realistically, however, a progressive Income Tax — or better yet a Wealth Tax — would be the best way to try to impose the cost burden for past pollution on those who benefited most from it. The other essential tool is the inheritance tax or levies on other transfers of assets from one party to another by deed or gift.

We are not concerned here with the precise form of taxation, but rather, with the equity and incentive implications of revenue-raising instruments. In the course of consideration of revisions to CERCLA, the US has conducted some of the most extensive studies of the costs involved in cleanups and the increased revenues required for different sorts of national state takeover of mitigation expenses. The potential tax increases are substantial. For example, the US case indicates that a state takeover of only the 'orphan shares' of cleanups for PRPs that cannot be identified or are no longer solvent (instead of distribution of those costs across other PRPs) would require a 71 per cent expansion in the magnitude of the Fund itself; full state takeover of liability for cleanup of sites with multiple polluters (to avoid the costly countersuits between PRPs) would require a 217 per cent revenue increase; full state responsibility for all sites on the National Priorities List as of 1992 would require an eightfold increase in revenues (Probst and Portney 1992).

6. *Contaminated land policy must also address assignment of liability for environmental damage and harm to human health in terms of incentives and equity impacts.* Both the EU and US clearly assign liability for the effects of current known pollution on the operators of facilities and on current landowners. Law and policy differs with respect to retroactive liability for past contamination, and with regard to prospective liability for newly discovered damage. Judicial rulings in the US ('Fleet Factors') and the UK ('Cambridge Water') take markedly opposite positions on liability for past pollution, exemplifying the distinction. The US assignment of liability to all parties in a chain of title is clearly intended to privatize the costs of reparations and repair (at whatever public or private transaction costs), while the finding in the UK recognizes a corporate responsibility for the damage done under the public or state standards that were in force when actions were taken.

The more the liability for past actions is privatized and potentially distributed across a large number of parties, the greater the investment risk associated with reclamation efforts. Privatization, which we have observed in the US, thus places a high burden on the (currently lower-income, possibly minority) areas that suffered the damage in the past, since private capital will, given alternatives, avoid those locales or those sites. Acceptance of some collective responsibility for past contamination could generate greater environmental equity, both by facilitating private capital flows to brownfield sites and by transferring state resources for cleanups to those areas adversely affected and placed at an economic development disadvantage, which is the objective of the EU regional funds.

However, the acceptance of a state share for liability returns us to the issue of tax policy. Given the vastly higher central jurisdiction expenditures that would be associated with higher levels of state responsibility, tax incidence, both geographically and across income groups, becomes a critical issue. Tax *equity*, not putative tax equality, thus has to be considered an environmental, as well as an economic, issue. In the absence of a tax system that is in some measure progressive with respect to both individual taxpayer wealth or income and spatial or regional wealth, state takeover of liability shares will not necessarily correct the inequality and injustice inevitably associated with the geographical concentration of heavy land contamination.

VARIETIES OF RISK EXPOSURE

As we argued in Chapter 1, all participants in an economy that is burdened with past contamination and abandoned or derelict land pay for urban blight. The cost each incurs is not clearly known or understood, and is pervaded with uncertainties. Thus each individual or firm within such economies is exposed to risks in some measure. We can, however, distinguish discrete groups on the basis of the voluntarism involved in their risk taking and the type of risk, financial, health or ecosystem, to which they are exposed. We first define our terms, then identify the different parties by their risk exposures, and third, turn to the effects of risk on behaviours.

Aspects of Risk Exposure

Voluntarism is a key factor: are risks willingly taken or are they imposed? Involuntary imposition of risks, especially if the burden is unequally distributed, heightens fears and reduces well-being. As Gilroy (1994:237) notes, 'There is a difference between private risk taking and collectively imposed risk'. Individual persons and communities may have risks imposed on them by private firms. Large economic institutions, firms and sub-national public sector entities may, in turn, have risks imposed on them by a central authority (the EU or US). Alternatively, as the German and French cases in Chapter 6 illustrate, the local or regional state can protect parties who might otherwise be deterred from investing in cleanup efforts from risk exposures.

Financial risk may be shared, in varying amounts by all parties. In fact, much of the litigation that comprises an apparently unacceptable portion of the total costs of the US Superfund programme is attributable to efforts to avoid such risks. Insurers' arguments that land contamination liabilities should not be covered by policies written before the passage of recent legislation are clear efforts at risk-burden shifting. As we have noted above, the uncertainty surrounding the magnitude of such risks actually may lead to their exaggeration.

Health risks, however, are problems only for living persons, as distinct from legal persons such as corporations, and, by extension, unincorporated businesses. The inadequacy of scientific knowledge of such risks explains some of the past contamination. Concern about advances in such knowledge is the source of risks of new liabilities for further cleanup under CERCLA. The distinguishing factor in such risks is that there can be no equitable distribution when risk exposures are caused by parties other than those experiencing the potential damage. Finally, such risks are generally assignable to and shared by a group of individuals with no direct legal links to the sources of the risks: the members of physically proximate communities.

Ecosystem risks present yet another problem. They may not be spatially concentrated, they may be cumulative, and they are not always easily identified in the short term. For our purposes here, although they constitute community concerns in many contexts, they are somewhat tangential, simply because spatial shifts in private investment or the allocation of severely limited public sector resources are not likely to be influenced by such long-term considerations, however much they should be.

These distinctions regarding the types of risk exposures lead us to a different enumeration of interested parties from the one we developed in Chapter 1 for those potentially damaged by harm to the common pool resource of the urban milieu. Obviously many households and organizations are affected both by blight and by contamination risks, but, as we discuss below, their motives to pursue policy change respond differently to the types of potential damage they confront. We distinguish here nine distinct parties exposed to risks associated with policies towards contaminated lands. (We leave out a number of possible groups, here, but the distinctions cover those that Scherr (1987:148) identified as the key actors, 'governments, industry, and citizens' organizations'.)

Lessons for Future Contaminated Land Policy: Prospects and Pitfalls 189

Parties Exposed to Risks

1. Current non-residential owners and users of the sites. To the extent that the on-site contamination was known to them, or they caused it, they are *voluntary* risk takers; if they are innocent landowners, their risks are involuntary. In either case, the risks they face are *financial*.

2. Past landowners and users of the sites. These comprise two sub-groups as well: those who contributed to the pollution and those who did not do so (with the latter including financial institutions and others who may have held fleeting title). However, all may be said to face *involuntary financial* risks in so far as their actions took place before new standards for avoiding or mitigating contamination came into play. A company which dumped toxics in accordance with accepted practice in the 1950s cannot be presumed to have, at the time, voluntarily accepted the risk of what is an altogether new form of regulation of land use.

3. Potential developers of the sites. These parties are faced with the decision of whether or not to incur *voluntary financial* risks. Instability in regulatory practices and standards, however, may produce levels of uncertainty that make it difficult for them to determine the extent of the risk they face.

4. Financiers backing potential site developers. These parties are analogous to the developers themselves, facing *voluntary financial* risks. However, they are less likely to enjoy potentially exceptional returns on their risk-taking behaviours (often lending at fixed interest rates), and thus tend to be more risk-averse.

5. Liability insurers, either for past contamination or for the firms engaged in current detection and mitigation. Insurers in the US have attempted to argue that they were exposed to involuntary financial risks in providing general business liability coverage which, after regulatory changes, was expected to cover claims for past land contamination that was not identified when the policies were written. However, in general, we can consider liability insurers to be in the business of risk-taking, and thus consider them to accept *voluntary financial* risks.

6. Firms engaged in toxics identification and removal from sites, hired by would-be developers or others. These 'expert' parties are exposed to *voluntary financial* risks in liability for both the quality of their analytical work and for toxic releases that may result from removal or mitigation efforts.

7. Regional or local public sector entities who, although not direct operators of sites (such as landfills), may have been waste generators and thus shared in *involuntary financial* risks, but who, in their economic development roles, may take on new *voluntary financial* risks as the lead agents in mitigation efforts, pursuing both new economic activity in their areas and reduction in toxics exposures for residents.

8. Residents on or near contaminated sites. Homeowners and other residents certainly faced *involuntary health* risks associated with past exposures to the contamination; to the extent that they pay their own healthcare costs, these parties would also suffer *involuntary financial* risks. Owner-occupants may have been exposed to *involuntary financial and ecosystem* risks as well, if their property values and/or landscaping were adversely affected by the nearby pollution.

9. Taxpayers at each level of government, who, while not involved in toxics disposal, may potentially become responsible for some costs of contamination mitigation. To the extent that businesses polluted in the past without informing the general public, mitigation costs can be considered to be involuntary financial burdens. However, to the extent that the potential cost burdens result from public pressure to clean up contaminated land, that is, to the extent that the new policies are expressions of political will, taxpayers face *voluntary financial* risks. On the other hand, in so far as taxpayers can be considered to be citizens of a country and as a group experience whatever negative consequences result from contamination, they have been, and remain, exposed to *involuntary health and ecosystem* risks.

The extent of risk exposure for any one party relates to the diversity of risks to which he or she is exposed. Depending on its risk exposure, each party that may play a role in reclaiming contaminated land will respond differently to similar imperatives or incentives. A brief look at

Aspects of Responses to Risk

Frank Knight launched modern economic analysis of financial risk in 1921. He argued that, for businesses, 'the two things, uncertainty-bearing and responsible control, are inseparable' (1921:350). To the extent that any public policy towards land contamination assigns responsibility for current and future cleanup costs retroactively on landowners, users, developers or financiers, the approach will tend to separate risk (his 'uncertainty') from control, thus imposing *involuntary financial* risks, which businesses eschew. The greater the retroactivity of the policy approach, the more it will tend to depress investment on brownfields.

The scale of contaminated land cleanup — and the scale of urban regeneration that is required in many depressed industrial cities — suggests that local reclamation efforts will have to attract capital from non-local source and international capital markets (Ryden 1992). Financial institutions with widespread investment, however, have been moving to centralize their risk assessment in efforts to coordinate policy and control risk exposures. MacCrimmon and Wehrung (1986) suggest that any such weakening of the authority of local managers tends to decrease the financial sector's willingness to accept even *voluntary financial* risk. Thus the very scale of the contamination problem in a locality will affect its ability to attract the private capital necessary for public or private parties to reclaim brownfield sites.

Community concerns over risks, largely derived from the fears of nearby residents, are driven by both fears of the unknown and the perception that the risks are increasing (United Nations Environment Programme 1992). The fact that their concerns include *financial, health and ecosystem* impacts not only increases the aggregate total risk, but also the uncertainty over risk levels. The imposition of *involuntary* risks on neighbourhoods contributes towards exaggeration of the dangers, as we indicate below. Timely and expeditious resident risk-minimization policies may, however, come into conflict with efforts to reduce the involuntary aspects of exposure through broadened local participation in decision-making on cleanups (Jasanoff 1986).

Taxpayers and citizens may want to reclaim contaminated lands. If they act on that intent, they tend to accept financial responsibility, at least as a matter of last resort. Policies that raise taxes to stimulate or pay for cleanups are thus clearly *voluntary* for taxpayers as a whole, even when the total costs are unknown and *financial* risk is present. The risk to the electorate at large is thus different from that to those residing or working near contaminated sites.

To the extent that the policies promulgated are perceived as failing to achieve their ends, recognition of *involuntary health and ecosystem* risks may grow. Dissatisfaction with the accomplishments of contaminated land policies may then lead either towards more enforcement and demands for cost-shifting to polluting firms and other parties or towards state takeover of the programmes, regardless of cost. The direction chosen may vary with the level of government and the broader context of policy, issues to which we return below.

Before we turn to the manner in which these different risk-exposed parties participate in shaping contaminated land policy, we need to digress briefly to discuss how the magnitude of a risk is determined. Policy analysts, cost–benefit analysts and statistical risk analysts all attempt to measure 'objective' risk. Their calculations typically combine the probability of an incident of a particular type occurring and the harm or benefit resulting if the event occurs. The product of the two measures produces a statistical risk measure, the 'expected' loss or gain.

This nominally objective measure, however, fails to describe how risks are actually perceived by those who must cope with them, whether they are households or businesses. Perceived, not 'objective' risk is the factor that motivates behaviours, pressures for policy approaches, and, in many ways, the condition of the common pool resource that is the urban ambience. We thus turn to a brief discussion derived from psychological and sociological literature to explicate investor and household perceptions of the risks involved in the reclamation and reuse of potentially contaminated brownfield properties.

Research in cognitive psychology has demonstrated the importance of addressing the subjective component of risk perception (Combs and Slovic 1979; Lichtenstein et al. 1978; Slovic 1979, 1992; Slovic et al. 1976). Individuals' (and organizations') evaluations of risk magnitude and frequency are sometimes inaccurate (e.g., estimates of the likelihood of hazardous events occurring do not correspond with actual statistical probabilities). Three types of characteristics which account for exceptional adversity and tendency towards overstatement are particularly relevant to brownfield reuse projects.

First, a risk is perceived to be more threatening and less acceptable when it is *unfamiliar*, and toxic chemicals provide the quintessential example: chemical pollutants *per se* have an especially menacing aura that evokes exceptional anxiety (Slovic 1992). Investors' own personal reactions may trigger their fears that, even when the property has been 'cleaned', doubts about toxic remnants will persist on the part of future renters and purchasers of the property.

Second, lack of control exacerbates risk aversion: the extent to which a situation evokes anxiety depends on the ratio of the degree of perceived threat to perceived abilities and resources to manage the threat (Folkman et al. 1979; Lazarus 1966). If individuals anticipate the ratio to be disadvantageous, they are likely to evade the threatening situation altogether. We have documented aspects of brownfield redevelopment projects which may be beyond the investors' or community residents' control, especially in the US context.

Third, issues of fairness shape conclusions regarding risk acceptability. Risks are perceived to be less tolerable when they are judged to be unjust and when they are imposed rather than voluntarily accepted (Hallman and Wandersman 1992; Slovic 1979; Starr 1969). These findings are highly pertinent to brownfield ventures: as we discussed above, costly liability obligations or unanticipated exposure to health threats may well be perceived as unfair.

All these factors tend to generate tendencies towards the exaggeration of risks, and to increasing aversion to them. They are compounded by the workings of the 'availability heuristic' (Kahneman and Tversky 1973, 1979; Lichtenstein et al. 1978). When people evaluate a particular risk, they rely on what they remember about similar risks. In doing so, they use inferential rules or heuristics which can lead to biased conclusions. The availability heuristic refers to the process of judging an event as likely or frequent if instances of it are easy to imagine or recall. This is the reason past experience with environmental policies and accidents has such an influence on policy: the negative events get publicity, and are thus the basis on which judgements are made. The wide publicity given NPL site cleanup efforts under CERCLA and the contentious litigation that has resulted have made the US particularly vulnerable to the workings of the availability heuristic. By contrast, the routinized, state-led cleanups in many EU member-states have held down the contentiousness of cleanup efforts, thus avoiding inappropriate exaggerations of the problems involved.

THE PRESSURES FOR POLICY CHANGES

The impetus for state intervention with respect to any issue derives from the social definition of the issue as a problem. The emergence of concerns for, and policies towards, contaminated land and its mitigation has been traced in our chapters on the US and EU. The political importance of any environmental problem is socially determined and thus a function of the success of different interest groups in generating activism. The interested actors, although they may not organize themselves along those lines, include two major classes of affected parties: first, the parties exposed to risk, whom we have just identified, and second, the actors affected by urban blight, whom we first considered in Chapter 1. Many parties are exposed to both risk and blight, but the two impacts act on them in different ways and can lead them to varying political postures. Thus we analyse them separately. We conclude with a look at other pressures, largely international in scope, which further help shape national US or supra-national EU policy.

Pressures from Parties Exposed to Risk

While there are many different parties exposed to risks by the existence of contaminated land and by policies towards mitigating contamination, it is probable that the most politically motivated groups are those incurring involuntary risks. The innocent landowners and users and the aggregates of people at local and regional levels, as well as all taxpayers, are forced to take on such risks. While the potential developers, their financiers and insurers accept risks voluntarily, they can control their exposure by how much work they do on brownfield sites. Since they tend to have more financial resources at their disposal than the others, if they perceive profit gains from policy changes, they may be able to influence the policy process with money power rather than people power. For the toxics management firms, contaminated land is their business; their concern is probably above all with the availability of insurance, since their workload and the fees they can charge both rise with concern for contamination.

Change is driven by experience of inappropriate stasis. Certainly the slow rate of Superfund cleanup completion discussed in Chapter 4 has been a cause of major dissatisfaction in the US, notwithstanding evidence that efforts and accomplishments are accelerating. No such all-encompassing legislation has been in place in Europe. The slow rate of progress at the EU level in passing Directives regarding contaminated land, not to mention the delays in the formation and activation of the European Environment Agency, described in Chapter 6, in itself, may be a source of dissatisfaction — and thus of pressure for change, *any* change.

In some measure, however, these problems of delay and the appearance of inaction derive directly from the diverse interests and concerns associated with the voluntary and involuntary risks we have identified. The sheer number of different interests and the extent to which they are in direct opposition to each other, especially as regards financial risks, increase the variety of political voices and, in so doing, slow both policy formation and implementation. It is probable that the slow pace evident in both the US and the EU is attributable in large part to the problems of forging compromise between these different interest groups. Thus we should not dwell overlong on the litigation that characterizes CERCLA. Similar conflicts confuse policy implementation and retard reclamation and redevelopment efforts in other contexts, but the disputes are not played out in higher visible courts of law.

Experience of unanticipated human health or ecological damage also stimulates pressures for policy changes. We have already traced the passage of CERCLA in part to the activism generated by a series of 'environmental disasters' in the US (although the evidence in retrospect raises questions about the damage actually done). Similarly in Europe, the Seveso Directive, which approximates the Title III 'Community Right-to-Know' provisions of the US's SARA, was driven by a major accident. Given the volume of chemicals still produced, transported, used in manufacturing and disposed of in various ways, it is statistically probable that further accidents and unintended releases will increase public pressure for cleanups from those fearing exposure to involuntary health, ecosystem and financial risks.

Pressures from Parties Affected by Blight

Another political motivation, distinct from that associated with risk, derives from the threat that land contamination poses to what we defined as the common pool resource we can call the city. Concern for lagging redevelopment of blighted urban centres and their growing social ills argues for policy change that will facilitate the reclamation and reuse of sites abandoned because of risks involving past contamination. To the extent that funding dilemmas cannot be resolved or cleanup costs not contained, these problems may generate pressure for relaxation of mitigation standards for past pollution.

Policy change in Europe is arguably driven by the need to create a 'level playing field' within the Single European Market (Cecchini 1988). Similar pressures exist within the US in terms of concern for differential cleanliness standards across the fifty states, and further developments under NAFTA will produce trans-national pressures of a similar variety (Mumme 1993; Reilly 1993). These are all forces arguing for some standardization, in order to avoid the creation of 'pollution havens'. Commonality of approach towards future actions that may damage the environment and to methods and standards for cleaning up past pollution, however, is not enough to assure equality of opportunity for development.

Unless full financial responsibility for the mitigation of contaminated lands is vested in the central jurisdiction of the common economic area, the EU in Europe or some yet non-existent supranational body in the case of North America, cleanup costs still accrue to the would-be users of individual brownfield sites. Thus communities and areas suffering from an overabundance of past contamination will be at a relative disadvantage in attracting investment. The EU response to such problems is its regional programmes, with special pools of funds set aside for redevelopment of old steel and coal regions and other depressed areas. No such targeted general revenue funds exist to level the playing field within the US, let alone across the vast expanse of land in North America as a whole.

To the extent that local areas must bear the costs of their past contamination, a real danger exists that, in some measure, the advanced industrial states (or the free trade blocs in which they operate) will declare a form of environmental bankruptcy. That is, they may conclude that the costs of cleanup exceed the benefits, and then declare the debt too great to be payable from the existing assets and cash flow of the nation-state. Such a response would have the same effect as a private sector bankruptcy: the lenders holding receipts payable will be left unpaid. Thus the urban centres which were the cradle of modern industry could be left to decay, with residents lacking the means or will to escape the health risks associated with continued contamination, shunned by capital that pursued greener pastures on greenfield sites.

Such a scenario has long seemed plausible in the US, with its massive land area and concentration of Superfund sites in the Northeast and Midwest states and California (and other major instances of land contamination concentrated in isolated mining areas). However, similar possibilities are arising in Europe as the Single European Market gains maturity. In the same way as population and development capital in the US has migrated between the fifty states, abandoning older industrial areas to fester with their contamination, the four freedoms are creating new opportunities for capital to abandon what have become peripheral industrial regions in favour of the so-called 'Blue Banana' — the new economically dominant core of Europe (Brunet 1989; Masser et al. 1991; Williams 1992).

Those suffering from the blight of their urban common pool resources, then, have strong motivations to pursue national or supranational cost-sharing and acceptance of responsibility. It is no accident that the successive efforts to further amend CERCLA since the 1980s have been spearheaded by legislators representing the old industrial belt in the Northeast and Midwest. Their constituents are the ones suffering the CPR losses and suffer the debilitation of their economic development prospects generated by the cleanup cost overburden, whichever locally-involved parties pay the bill. From a locally parochial perspective, it does not matter much if local firms or a municipality itself has to pay the bill for reclamation of contamination. In either case, the capital committed to mitigation is not available for other purposes which benefit the urban CPR and it becomes difficult to attract new capital, which would either have to pay higher development costs for reclaiming brownfield sites or higher taxes to the local government if it needed to finance cleanups it undertook itself.

The very existence of spatial environmental inequality, which is an inevitable consequence of the geographic concentration of manufacturing in the past, creates strong pressures in the direction of implementation of some of the guiding principles for contaminated land policy which we have identified. The concentration of racial and ethnic minority groups in the adversely impacted locales, especially marked in the case of the US but also evident in portions of Europe, accentuates this problem of environmental justice. The role this factor will actually play in shaping environmental policy will depend on the extent to which race- or ethnicity-based discriminatory attitudes influence political discourse. Given the recent successes of right-wing political forces in many of the nations in question, this factor could, unfortunately, impede the advance of environmental justice and systematic reclamation and redevelopment of contaminated brownfield land.

Other Pressures

As we have seen, the mechanisms in place for addressing what Barnett, writing in 1994, described as the US *Toxic Debt* are inadequate by any measure of debt service. New sites are being uncovered far more rapidly than old ones are being successfully mitigated. The non-productive transaction costs associated with addressing the existing sites seem to be rising everywhere, although driven to exceptional heights by the litigiousness that characterizes the US approach. Community concern is not being adequately dispelled even when treatments are applied to mitigate contamination, and, as we have also seen, policy uncertainty and disarray are evident at the level of the nation-state and above on both sides of the Atlantic.

The broad elements of approaches to discharging the debts accumulated by past contamination have been obvious for a long time. In the ideal, methods should be devised that minimize adverse risk consequences, distribute financial burdens both equitably and reasonably efficiently, reduce human health and ecosystem damage by prompt action, provide incentives for reduced future contamination, and correct for the distortions in capital flows which have contributed to urban decay in the advanced industrial states of Western Europe and North America. These pressures do not abate with time. Rather, like the total indebtedness on a loan on which no payments are made, they compound; the capacity of the nation-state to service the debt without massive sacrifice of other economic activity may decline.

The expansion of the EU and growing preponderance of member-states with strong environmental standards may result in final approval of environmental Directives with the scope of RCRA and CERCLA in the US. Such policies may be implementable without a replication of the US capital flight experience. The EU regional policy concern for overcoming regional disparities and intervening to assist economically backward and peripheral regions is central to such an adaptation. The network of special funds targeted to regions experiencing economic downturns due to economic restructuring are intended to mitigate the damage done by private capital outflows. More important, perhaps, because of its influence on the perceived risks of key parties, is European corporatism, which both reduces costs to individual firms and has resulted in a higher level of trust of the state as an economic actor than is the norm in the US.

The evolution of EU environmental policy towards contaminated land reclamation will, in turn, influence that prevailing in the US. Paul

Portney of Resources for the Future is typical of American commentators in his expressed fear (1992:17) 'that those [firms] subject to US hazardous waste regulations find themselves increasingly in competition with foreign manufacturers that face less stringent regulatory standards'. His comment fails to distinguish between standards for current behaviour (which are certainly higher in practice in many parts of Europe than in North America), and imposition of cost burdens for past behaviours (which are heavier in the US). In focusing on manufacturers, moreover, he fails to address the broad issue of capital flows to an array of different business sectors.

An EU policy that recognizes collective responsibility for past pollution even while adhering to a polluter pays principle for current production, transport, utilization and disposal of wastes, especially toxics, might be expected to influence the evolution of US practice. The 1994 US congressional election, however, suggests the opposite, that Europe and the US may be headed in different directions. The Republicans gained control of both Houses of Congress with calls for reduction in the range of powers and budget resources of the nation-state in favour of the fifty individual units. Moreover, NAFTA and GATT both were ratified only after the individual states got reassurances about their rights to retain certain local controls on labour and environmental practices. The American individualist ethos appears relatively unable to detect a 'divide and conquer' spatial location or investment strategy on the part of trans-national (or, in the US case, trans-state) capital and corporations. European corporatism, by contrast, not only recognizes the strategy, but has greater faith in the mechanisms of the state, and supra-national aggregates, to respond and minimize negative consequences.

CONCLUSIONS: WHAT WORKS?

No single response to the question of the optimal contaminated land policy is possible given the different contexts prevailing on the two sides of the Atlantic. Yet some lessons may be extracted from our comparative efforts.

In General

We can conclude from our cases and other evidence that neither nationalization nor privatization of the urban Common Pool Resource that is affected by contamination can be relied upon. The standards for cleanliness and the means of contamination mitigation *have* to be sensitive to local land markets and community concerns if they are to both satisfy residents and gain local acceptance (thereby avoiding deterioration of land values and other aspects of the CPR) and satisfy non-local private sources of capital that are essential to redevelopment of brownfield sites.

Decentralized collective management of some form is thus the only viable approach to preservation and regeneration of the urban CPR. Such an approach, in turn, dictates very specific roles for the overarching central jurisdiction that sets the regulatory standards and assigns legal responsibilities. We have alluded to the fact that local standards can be undermined by the economic desperation of areas with a surfeit of possibly contaminated brownfield sites, since this condition may be accompanied by lower than average income levels or levels of economic opportunity. We have also noted that, in both the EU and the US, the central jurisdiction has responded to the demand of member-states for the right to impose higher standards of cleanliness than the centrally-dictated norm. Three inter-locking policies or practices at the central jurisdiction level thus seem imperative if decentralized collective management is to be attained.

1. *A high centrally-defined standard of cleanliness to which all local jurisdictions must conform.* This standard is necessary to avoid any local sacrifice of the environment for economic development and to provide a level playing field. The right to supersede the standard with more stringent conditions should be available, but those higher levels of cleanup should be locally financed (as is the case under both EU and US current practice).

2. *Centrally-financed redistributive pools to assist those localities suffering from a geographic concentration of toxic debts that need to be discharged — and to have the option of imposing higher than minimum standards if they so desire.* Such funds are basic to the provision of environmental, not just economic, justice. In the EU, where such funds exist, pursuit of a cleaner environment is seen as a complement to economic development efforts; in the US, where they are not present, environmental and economic development goals are perceived as conflicting objectives.

3. *Acceptance by the central jurisdiction of responsibility for any liability that may arise in the future from inadequate mitigation of contamination in the present.* The issue of prospective liability arising from new discoveries or detection techniques has received minimal attention in the European context, although the spectre of the contaminated zones of Central and Eastern Europe bring the problem into high relief. In the US, such liability is assigned to the CERCLA-designated Potentially Responsible Parties, with devastating impacts on rates of non-federally-financed reclamation and reuse efforts. In both settings, however, such public acceptance of liability would serve the other two policy imperatives while assuring consistency and predictability in markets in the future. The liability exposure for the central jurisdiction would increase the pressure for implementation of high, rather than low, basic mitigation standards. Assuring the private and sub-national public sectors that they would not face such prospective liabilities would also reduce the negative effects of existing contamination on land development investment projections, thus reducing the magnitude of the redistributive pools required to level the playing field.

For the European Union

The US practice of strict liability assignment to private parties for past pollution is a far more viable tactic than observation of the current CERCLA experience might suggest. The existence of stricter land-use controls, restrictions on development, greenbelts, and the generally greater scarcity of urbanizable or developable greenfield land all reduce the options to brownfield redevelopment that are available to investors, thus lowering the competitive disadvantage of potentially polluted sites that carry liability. Assuming the general principles enumerated above are adhered to, there appears to be relatively little risk of acute land market distortion associated with the imposition of a more stringent form of the polluter pays principle.

For the United States

Comparison of the US experience with that of the nations of Europe underscores the basic flaw in US policy, one that is not a matter of law but of policy ethos. So long as a strict individualist ethos prevails, no tinkering at the margins of CERCLA or other contamination mitigation laws and regulations will significantly improve conditions and the speed and quality of cleanups. We have argued and demonstrated that a toxic debt has been accumulated by the society and economy as a whole; until and unless the retirement of that collective debt is recognized as requiring a corporatist or collective response, inefficient practices and tendencies towards the exacerbation of environmental and economic inequality will remain the norm.

References

AAV (1993), *Jahresbericht des Abfallentsorgungs-und Altastentaniersverbandes*, Nordrein-Westfalen: Hattingen.
Acosta, R. and V. Renard (1993), *Urban Land and Property Markets in France*, London: University College – London Press.
Acton, J.P. and P.C. Beider (1992), 'Superfund Liability Provisions: Impact on the Industrial and Insurance Sectors', in *U.S. Waste Management Policies: Impact on Economic Growth and Investment Strategies*, American Council for Capital Formation, Center for Policy Research (eds), Washington, DC: American Council for Capital Formation, Center for Policy Research, pp. 70–85.
Association des Régions Européennes de Technologie Industrielle (RETI), (1992), *Study of Derelict Land in the ECSC*, Volume A: Summary and Recommendations, Brussels: RETI.
Baes, C.F. and G. Marland (1989), *Evaluation of Cleanup Levels for Remedial Action at CERCLA Sites Based on a Review of EPA Records of Decision*, Oak Ridge, TN: Oak Ridge National Laboratories.
Barnekov, T., R. Boyle and D. Rich (1989), *Privatism and Urban Policy in Britain and the United States*, New York: Oxford University Press.
Barnett, H.C. (1994), *Toxic Debts and the Superfund Dilemma*, Chapel Hill, NC: University of North Carolina Press.
Bartsch, C., C. Andress, J. Seitzman and D. Cooney (1991), *New Life For Old Buildings: Confronting Environmental and Economic Issues to Industrial Reuse*, Washington, DC: Northeast-Midwest Institute.
Bartsch, C. and E.E. Collaton (1994), *Industrial Site Reuse, Contamination, and Urban Redevelopment: Coping With the Challenges of Brownfields*, Washington, DC: Northeast Midwest Institute.
Baum, A., R. Fleming and J.E. Singer (1983), 'Coping with victimization by technological disaster', *Journal of Social Issues* **XXXIX** (2): 117–18.
Bergman, E.M. (ed.) (1986), *Local Economies in Transition*, Durham, NC: Duke University Press.
Beyer, S. (1982), *Regulation and its Reform*, Cambridge, MA: Harvard University Press.
Bingham, R.D. and J.P. Blair (eds) (1984), *Urban Economic Development*, Beverly Hills, CA: Sage Publications.
Bingham, R.D. and R. Mier (eds) (1993), *Theories of Local Economic Development*, Newbury Park, CA: Sage Publications.
Birch, D.L. (1970), *The Economic Future of City and Suburb*, Supplementary Paper Number 30, New York: Committee for Economic Development.
Bloom, G. (1986), 'The hidden liability of hazardous waste cleanup', *Technology Review* **LXXXIX**: 58–65.

Boman, C.R. (1991), 'Creative acquisitions of contaminated property', *Probate and Property* January: 28–31.
Bradbury, J.A. (1989), 'The policy implications of differing concepts of risk', *Science, Technology, and Human Values* **XIV**: 380–99.
Brittan, Sir L. (1992), Speech to conference on the Maastricht Treaty, University of Hull, May.
Brown, M. (1981), *Laying Waste: the Poisoning of America by Toxic Chemicals*, New York: Washington Square Press.
Brown, M.S. (1987), 'Communicating Information about Workplace Hazards: Effects on Worker Attitudes Toward Risks', in B.B. Johnson and V.T. Covello (eds), *The Social and Cultural Construction of Risk: Essays on Risk Selection and Perception*, Boston: D. Reidel, pp. 252–74.
Brown, P. and E.J. Mikkelsen (1990), *No Safe Place: Toxic Waste, Leukemia, and Community Action*, Berkeley: University of California Press.
Brundtland Commission (World Commission on Environment and Development) (1987), *Our Common Future*, London: Oxford University Press.
Brunet, R. (1989), *Les Villes Européennes*, Paris: DATAR, La Documentation Française.
Bryant, B. and P. Mohai (eds) (1992), *Race and the Incidence of Environmental Hazards; A Time for Discourse*, Boulder, CO: Westview Press.
Buchert, R.J. (1990), Statement for the Record, in US House of Representatives, Committee on Energy and Commerce, Subcommittee on Transportation and Hazardous Materials, *Hearings on H.R. 4494: Lender Liability Under Superfund*, 2 August, Washington, DC: USGPO, pp. 194–60.
Budd, S. and A. Jones (1991), *The European Community: A Guide to the Maze*, Fourth Edition, London: Kogan Page.
Bullard, R.D. (1990), *Dumping in Dixie: Race, Class and Environmental Quality*, Boulder, CO: Westview Press.
Bullard, R.D. (ed.) (1993), *Confronting Environmental Racism: Voices from the Grassroots*, Boston, MA: South End Press.
Bullard, R.D. and B. H. Wright (1992), 'The Quest for Environmental Equity: Mobilizing the African-American Community for Social Change', in *American Environmentalism: The U.S. Environmental Movement, 1970–1990*, R.E. Dunlap and A.G. Mertig (eds), Philadelphia, PA: Taylor and Francis, pp. 11–26.
Cambridge Reports/Research International (1990), *The green revolution and the changing American consumer*, Cambridge, MA: Cambridge Reports/Research International.
Cambridge Reports, Inc. (1986), 'Paying for environmental quality', *Bulletin on Consumer Opinion*, No 112. Cambridge, MA: Cambridge Reports Inc.
Campbell, J.M. (1992), 'Lender Liability for Environmental Cleanup: Effect on the Financial Services Industry', in American Council for Capital Formation, Center for Policy Research (eds), *U.S. Waste Management Policies: Impact on Economic Growth and Investment Strategies*, Washington, DC: American Council for Capital Formation, Center for Policy Research, pp. 45–60.
Carson, Rachel (1962), *Silent Spring*, New York: Houghton-Mifflin.
Cecchini, P. (1988), *The European Challenge: The Benefits of a Single Market*, Aldershot, UK: Wildwood House.

References

Chakravorty, S. (1992), 'Commercial Aspects of Section 143 of the Environmental Protection Act', *Journal of Planning and Environmental Law*, July: 624–7.

Church, T.W. and R.T. Nakamura (1993), *Cleaning Up the Mess: Implementation Strategies in Superfund*, Washington, DC: The Brookings Institute.

Colla, E.W. (1991), Statement for the Record, in US Senate, Committee on Banking, Housing, and Urban Affairs, *Hearings on Environmental Liability Issues*, 12 June, Washington, DC: USGPO, pp. 99–109.

Combs, B. and P. Slovic (1979), 'Newspaper coverage of causes of death', *Journalism Quarterly* **LVI**: 837–49.

Commission of the European Communities (CEC) (1989), *Proposal for a Council Directive on Civil Liability for Damage Caused by Waste*, COM (89) 282, Brussels, BE: CEC.

CEC (1990), *Green Paper on the Urban Environment*, COM (90) 218 final, Brussels, BE: CEC.

CEC (1991), 'Amended Proposal for a Council Directive on Civil Liability for Damage Caused by Waste', COM (91) 219, Brussels, BE: CEC.

CEC (1992a), *Fifth Action Programme on the Environment — Towards Sustainability*, COM (92) 23 final, Brussels, BE: CEC.

CEC (1992b), *The State of the Environment in the European Community*, COM (92) 23 final, Vol. III, Brussels, BE: CEC.

CEC (1993a), *Green Paper on Remedying Environmental Damage*, COM (93) 47 final, Brussels, BE: CEC.

CEC (1993b), *White Paper on Growth, Competition and Employment*, COM (93) 700, Brussels, BE: CEC.

Confederation of British Industries (1993), *CBI's Response to the EC's Green Paper on Remedying Environmental Damage, COM (93) 47*, London: Confederation of British Industries.

Connell, C. (1994), *Liability for Contaminated Land in Germany, France, Netherlands and Denmark*, London: Institute for European Environmental Policy.

Council of Europe (CE) (1993), *Convention on Civil Liability for Damage resulting from activities dangerous to the environment*, European Treaty Series 150, 21 June, Lugano, SW: CE.

Cuyahoga County Planning Commission (1992), *Brownfield Reuse Strategies: A Local Development Symposium*, Cleveland, OH: Cuyahoga County Planning Commission.

Davies, H.W.E. (1989), *Planning Control in Western Europe*, London: HMSO.

de Michaelis, G. (1990), Speech as President of the European Council, 'The Programme to Implement the Single European Market', Speech, Rome, Italy.

Devall, B. and G. Sessions (1985), *Deep Ecology: Living as if Nature Mattered*, Layton, UT: Gibbs M. Smith.

Dieterich, H. and E. Dransfeld (1992), 'Germany', in B. Wood and R.H. Williams (eds), *Industrial Property Markets in Western Europe*, London: E&FN Spon, Chapter 3.

Dieterich, H., E. Dransfeld and W. Voss (1993), *Urban Land and Property Markets in Germany*, London: University College – London Press.

Dieterich, H., R.H. Williams and B. Wood (1993–94), *European Urban Land and Property Markets*, Six Volumes, London: University College – London Press.

Dixon, L.S., D.S. Drezner and J.K. Hammitt (1993), *Private-Sector Cleanup Expenditures and Transaction Costs at 18 Superfund Sites*, Santa Monica, CA: RAND Corporation.

Douglas, M. and A. Wildavsky (1982), *Risk and Culture: An Essay on the Selection of Technological and Environmental Dangers*, Berkeley, CA: University of California Press.

Dransfeld, E. and W. Voss (1993), *Funktionsweise städtischer Bodenmärkte in Mitgliedstaaten der Europäischen Gemeinschaft — ein Systemvergleich*, Bonn: Bundesbauministerium.

Dror, Y. (1971), *Ventures in Policy Sciences*, New York: Elsevier.

Dunlap, R.E. (1992), 'Trends in Public Opinion Toward Environmental Issues: 1965–1990', in R.E. Dunlap and A.G. Mertig (eds), *American Environmentalism: The U.S. Environmental Movement, 1970–1990*, Philadelphia, PA: Taylor & Francis, pp. 89–116.

Dunlap, R.E. and A.G. Mertig (eds) (1992), *American Environmentalism: The U.S. Environmental Movement, 1970–1990*, Philadelphia, PA: Taylor & Francis.

Dye, T.R. (1966), *Politics, Economics, and the Public*, Chicago: Rand-McNally.

Eaton, L. (1994), 'New tests at "poison" nursery', *Independent* London, 1 September, p. 6.

Economist (1994), 'Where bankers fear to tread', 21 May: 89–90.

Edelstein, M. (1988), *Contaminated Communities: The Social and Psychological Impacts of Residential Toxic Exposure*, Boulder, CO: Westview.

Edison Electric Institute (1988), *How Clean is Clean? Clean-up Standards for Groundwater Soil*, Washington, DC: Edison Electric Institute.

Edulgee, G. (1994), 'Policy Profile: European Union Waste Policy', *European Environment* **IV** (3): 18–20.

Environmental Defense Fund (1994), 'Dioxin risks found greater than had been thought', *EDF Letter* **XXV** (6): 1, 3.

Erikson, K. (1990), 'Toxic Reckoning: Business Faces a New Kind of Fear', *Harvard Business Review* **LXVIII**: 188–26.

European Community Council (1992), *Treaty of European Union*, Luxembourg: OOPEC.

European Parliament, Committee on the Environment, Public Health and Consumer Protection (1991), *REPORT on the urban environment*, PE 150.274/fin, 6 August, Brussels, BE: European Parliament.

Fairclough, A. (1983), 'The Community's Environment Policy', in R. Macrory (ed.), *Britain, Europe and the Environment*, London: Imperial College Centre for Environmental Technology, pp. 19–34.

Faludi, A. (ed.) (1993), *Dutch Strategic Planning in International Perspective*, Amsterdam: SISWO.

Finsterbusch, K. (1989), 'Community responses to exposure to hazardous wastes', in D.L. Peck (ed.), *Psychosocial Effects of Hazardous Toxic Waste Disposal on Communities*, Springfield, IL: Charles C. Thomas, pp. 57–79.

Fischhoff, B., P. Slovic, S. Lichtenstein, S. Read and B. Combs (1978), 'How Safe is Safe Enough? A Psychometric Study of Attitudes Towards Technological Risks and Benefits', *Policy Sciences* **IX**: 127–52.
Fogleman, V.M. (1992), *Hazardous Waste Cleanup, Liability, and Litigation: A Comprehensive Guide to Superfund Law*, Westport, CT: Quorum.
Folkman, S., C. Schaefer and R.S. Lazarus (1979), 'Cognitive Processes as Mediators of Stress and Coping', in V. Hamilton and D. Warburton (eds), *Human Stress and Cognition: An Information Processing Approach*, New York: John Wiley & Sons, pp. 265–98.
Forte, Joseph P. (1991), 'Statement on Behalf of the American Bar Association', in United States Congress, Senate Committee on Banking, Housing, and Urban Affairs, *Hearings on Environmental Lender Liability Issues...*, 12 June, Washington, DC: USGPO, pp. 155–87.
Franzen, R. (1994), 'Oregon's Takings Tangle', *Planning*, **LX** (6): 13–15.
Free, J.H. (1993), 'Risk Perceptions Associated with a High-Level Nuclear Waste Repository', *Sociological Spectrum* **XIII**: 139–51.
Gamson, W. and A. Modigliani (1989), 'Media Discourse and Public Opinion on Nuclear Power: A Constructionist Approach', *American Journal of Sociology* **LXXXXV**: 1–37.
Gandy, M. (1993), Environmental Policy in Transition: A Comparative Overview of Key Developments in Britain and Germany, in J. Holder et al. (eds), *Perspectives on the Environment*, Aldershot, UK: Avebury, pp. 119–28.
Gedicks, A. (1993), *The New Resource Wars: Native and Environmental Struggles Against Multinational Corporations*, Boston, MA: South End Press.
Genske, D.D. and H-P. Noll (1994), 'Reclaiming Abandoned Mining Sites — Reorganization Concepts and Examples', Paper presented at the International Land Reclamation and Minerage Conference, Pittsburgh, PA, April.
Gilroy, J.M. (1994), 'Public Policy and Environmental Risk: Political Theory, Human Agency, and the Imprisoned Rider', *Environmental Ethics* **XIV**, Fall: 217–37.
Glaser, M. (1994), 'Economic and Environmental Repair in the Shadow of Superfund: Local Government Leadership in Building Strategic Partnerships', *Economic Development Quarterly* **VIII** (4): 345–52.
Glickman, J. and M. Judy (1993), 'Managing Superfund's Impact: A First Aid Kit', *Management Information Service* **25** (1), Washington, DC: International City Management Association.
Goldsteen, J.B. (1993), *Danger All Around: Waste Storage Crisis on the Texas and Louisiana Gulf Coast*, Austin, TX: University of Texas Press.
Goodman, R. (1979), *The Last Entrepreneurs: America's Regional Wars for Jobs and Dollars*, Boston: South End Press.
Gordon, J. (1994), 'Environment Policy in Britain and Germany: Some Comparisons', *European Environment* **IV** (3): 9–12.
Greve, M.S. and F.L. Smith (eds) (1992), *Environmental Politics: Public Costs, Private Rewards*, New York, NY: Praeger Publishers.
Gupta, S., G.V. Houtven and M.L. Cropper (1994), *The Value of Superfund Cleanups: Evidence from U.S. Environmental Protection Agency Decisions*, Washington, DC: The World Bank Environment Department Pollution and

Environmental Economics Division and Policy Research Department, Environment, Infrastructure, and Agriculture Division.
Hadden, S.G. (1991), 'Public Perceptions of Hazardous Waste', *Risk Analysis* **XI**: 47–57.
Haigh, N. (1989), *EEC Environment Policy and Britain*, Second Edition, Harlow: Longman.
Halliday, I., G. Marcou and R. Vickerman (1991), *The Channel Tunnel. Public Policy, Regional Development and European Integration*, London: Belhaven.
Hallman, W.K. and A. Wandersman (1992), 'Attribution of Responsibility and Individual and Collective Coping with Environmental Threats', *Journal of Social Issues* **XXXXVIII** (4): 101–18.
Hardin, G. (1968), The Tragedy of the Commons, *Science* **CLXVII**: 1243–8.
Harris, J. (1992), 'The Environmental Protection Act 1990 — Penalising the Polluter', *Journal of Planning and Environmental Law*, June: 515–24.
Healey, P. and R. Nabarro (1990), *Land and Property Development in a Changing Context*, Aldershot, Hants, UK: Gower.
Healey, P. and R.H. Williams (1993), 'European Urban Planning Systems: Diversity and Convergence', *Urban Studies* **XXX** (4/5): 701–20.
Helvarg, D.B. (1994), *The War Against the Greens: The 'Wise Us', Movement, The New Right, and Anti-Environmental Violence*, San Francisco, CA: Sierra Club Books.
Hennings, G. and K.R. Kunzmann (1993), 'Local Economic Development in a Traditional Industrial Area: The Case of the Ruhrgebiet', in P.B. Meyer (ed.), *Comparative Studies in Local Economic Development: Problems in Policy Implementation*, Westport, CT: Greenwood Press, pp. 35–54.
Hird, J.A. (1994), *Superfund: The Political Economy of Environmental Risk*, Baltimore, MD: Johns Hopkins University Press.
Hofferbert, R.I. (1974), *The Study of Public Policy*, Indianapolis, IN: Bobbs-Merrill.
Hofrichter, R. (ed.) (1993), *Toxic Struggles: the Theory and Practice of Environmental Justice*, Philadelphia: New Society Publishers.
Holland, M. (1992), Presentation to Brownfields Reuse Strategies: A Local Development Symposium, 10 October, Cited in Cuyahoga County (Ohio) Planning Commission. 1992. *Summary of the Proceedings*, Cleveland, OH: Cuyahoga Planning Commission, p. 57.
Jacobs, H.M. (1993), '"Wise Use" versus "The New Federalism": Social Conflict over Property Rights', Paper presented at the annual meeting of the Association of Collegiate Schools of Planning, Philadelphia, PA, October.
Jacobs, H.M. (1994), 'Rethinking progressive Localism? The "County Movement" for Conservative Control of Land Use Planning', Paper presented at the annual meeting of the Association of Collegiate Schools of Planning, Phoenix, AR, October.
Jasanoff, S. (1986), *Risk Management and Political Culture: A Comparative Study of Science in the Policy Context*, New York: Russell Sage Foundation.
Johnson, B.B. and V.T. Covello (1987), *The Social Construction of Risk*, Boston: D. Reidel.
Johnson, F.R., A. Fisher, V.K. Smith and W.H. Desvousges (1988), 'Informed Choice or Regulated Risk? Lessons from a Study in Radon Risk Communication', *Environment* **XXX** (4): 12–15, 30–35.

Just, R.E., D.L. Hueth and A. Schmitz (1982), *Applied Welfare Economics and Public Policy*, Englewood Cliffs, NJ: Prentice Hall.
Kahneman, D. and A. Tversky (1973), 'Availability: A Heuristic for judging frequency and probability', *Cognitive Psychology* **IV**: 207–32.
Kahneman, D. and A. Tversky (1979), 'Prospect theory: Analysis of decision under risk', *Econometrica* **XXXXVII**: 263–91.
Kalbro, T. and H. Mattsson (1995), *Urban Land and Property Markets in Sweden*, London: University College – London Press.
Kasperson, R.E. (1992), 'The social amplification of risk: Progress in developing an integrative framework', in S. Krimsky and D. Golding (eds), *Social Theories of Risk*, Westport, CT: Praeger, pp. 153–78.
Kasperson, R.E., D. Golding and S. Tuler (1992), 'Social Distrust as a Factor in Siting Hazardous Facilities and Communicating Risk', *Journal of Social Issues* **XXXXVIII** (4): 161–87.
Kasperson, R.E. and P.J.M. Stallen (eds), (1991), *Communicating Risks to the Public*, Boston: Kluwer.
Keene, J.C. (1993), 'An Approach to Determining Whether a Public Regulation Has "Taken" Private Property without Just Compensation', Paper presented at the annual meeting of the Association of Collegiate Schools of Planning, Philadelphia, PA, October.
Kelley, E.W. (1991), Statement for the Record, in US Senate, Committee on Banking, Housing, and Urban Affairs, *Hearings on Environmental Liability Issues*, 12 June, Washington, DC: USGPO, pp. 99–109.
Kinnard, W.N. (1989), 'Analyzing the Stigma Effect of Proximity to a Hazardous Materials Site', *Environmental Watch* December: 4–7.
Kramer, H. (1993), 'The EC's Response to the 'New' Eastern Europe', *Journal of Common Market Studies* **XXXI** (2): 213–44.
Krimsky, S. and A. Plough (1988), *Environmental Hazards: Communicating Risks as a Social Process*, Dover, MA: Auburn House.
Kunreuther, H. (1992), 'A Conceptual Framework for Managing Low-Probability Events', in S. Krimsky and D. Golding (eds), *Social Theories of Risk*, Westport, CT: Praeger, pp. 301–20.
Kunzmann, K.R. (1992), 'Zur Entwicklung der Stadtsysteme in Europa Mitteilungen der Österreichischen', *Geographischen Gesellschaft* **CXXXIV**: 25–50.
Kunzmann, K.R. and M. Wegener (1991), *The Pattern of Urbanisation in Western Europe*, IRPUD Berichte 28, Dortmund: Fakultät Raumplanung, Universität Dortmund.
Laconte, P. (1992), *Perspectives on Planning and Urban Development in Belgium*, Dordrecht, BE: Kluwer.
LaFalce, John (1989), 'Opening Remarks', in US Congress, House of Representatives, Committee on Small Business, *Hearings: Lender Liability under Superfund*, 3 August, Washington, DC: USGPO, pp. 1–3.
Landy, M.K., M.J. Roberts and S.R. Thomas (1994), *The Environmental Protection Agency: Asking the Wrong Questions From Nixon to Clinton*, New York: Oxford University Press.
Larson, J.S. (1992), 'Commentary on P.R. Portney's "The Economics of Hazardous Waste Regulation"', in American Council for Capital Formation, Center for

Policy Research (eds), *U.S. Waste Management Policies: Impact on Economic Growth and Investment Strategies*, Washington, DC: American Council for Capital Formation, Center for Policy Research, pp. 9–31.

Lasswell, H.D. (1971), *A Preview of Policy Sciences*, New York: Elsevier.

Lautenberg, F. (1993), 'Prepared Statement', in US Senate, Committee on Environment and Public Works, Subcommittee on Superfund, Recycling, and Solid Waste Management, *Hearings on Superfund Reauthorization*, 6 May, Washington, DC: USGPO, pp. 3–5.

Lazarus, R.S. (1966), *Psychological Stress and the Coping Process*, New York: McGraw-Hill.

Leopold, A. (1949) (1977), *A Sand Country Almanac: And Sketches Here and There*, New York: Oxford University Press.

Levine, A. and G. Levine (1982), *Love Canal: Science, Politics, and People*, Lexington, MA: Lexington Books.

Lewis, J. (1991), 'Superfund, RCRA, and UST: The Clean-up Threesome', *EPA Journal*, July/August: 29–33.

Lichtenberg, J. and D. MacLean (1991), 'The Role of the Media in Risk Communication', in R.E. Kasperson and P.J.M. Stallen (eds), *Communicating Risks to the Public*, Boston: Kluwer, pp. 157–74.

Lichtenstein, S., P. Slovic, B. Fischoff, M. Layman and B. Combs (1978), 'Judged Frequency of Lethal Events', *Journal of Experimental Psychology: Human Learning and Memory* **IV** (6): 551–78.

Light, C.R. (1990), 'Easing the Lender's Plight under CERCLA: A Risk Management Program', *Environmental Law Journal of Ohio* **2** (2): 241–3.

Lodge, J. (ed.) (1993), *The European Community and the Challenge of the Future*, Second Edition, London: Pinter.

Lowi, T.J. (1969), *The End of Liberalism: Ideology, Policy and the Crisis of Public Authority*, New York: W.W. Norton & Co., Inc.

Lynn, F.M. (1987), 'OSHA's Carcinogens Standard: Round One on Risk Assessment Models and Assumptions', in B.B. Johnson and V.T. Covello (eds), *The Social and Cultural Construction of Risk*, Boston: D. Reidelpp 345–58.

MacCrimmon, K.R. and D.A. Wehrung (1986), *Taking Risks: The Management of Uncertainty*, New York: The Free Press.

Mangun, W.R. (1988), 'A Comparative Analysis of Hazardous Waste Management Policy in Western Europe', in C.E. Davis and J.P. Lester (eds), *Dimensions of Hazardous Waste Politics and Policy*, New York: Greenwood Press, pp. 205–21.

Marshall, T. (1993), 'Regional Environmental Planning: Progress and Possibilities in Western Europe', *European Planning Studies* **I** (1): 69–90.

Martin, K.M. (1991a), 'Public/Private Cooperation in the Development of Contaminated Properties (part 1)', *Real Estate/Environmental Liability News* **II** (15): 9–12.

Martin, K.M. (1991b), 'Public/Private Cooperation in the Development of Contaminated Properties (part 2', *Real Estate/Environmental Liability News* **II** (16): 8–12.

Masser, I., O. Sviden, and M. Wegener (1992), *The Geography of Europe's Futures*, London: Belhaven.

References

Mazur, A. (1991), 'Putting Radon and Love Canal on the Public Agenda', in R. Couch and J.S. Kroll-Smith (eds), *Communities At Risk: Collective Responses to Technological Hazards*, New York: Peter Lang, pp. 183–204.

Meyer, P.B. (1991), 'Problem Recognition and Planning Response: Local Economic Development Practices in the U.S. and U.K.', Paper presented at the July joint meetings of the Association of Collegiate Schools of Planning and the Association of European Schools of Planning, Oxford, UK.

Meyer, P.B. (1993), 'Shaping Place: Institutional Behaviors and their Motivations', *Journal of Urban Affairs* **15** (6): 413–20.

Meyer, P.B. (1994), *Commentary on the 'Superfund Reform Act of 1994'*, Paper prepared at the request of the Senior Counsel, US Senate Banking Committee, Louisville, KY: Center for Environmental Management, University of Louisville.

Meyer, P.B. and M.A. Burayidi (1992), 'Arcadian Reinventions? Garden Festivals in Older British Industrial Cities', paper presented at the Urban Affairs Association meeting, April.

Mishan, E.J. (1981), *Introduction to Normative Economics*, New York: Oxford University Press.

Mitchell, R.C., A.G. Mertig and R.E. Dunlap (1992), 'Twenty Years of Environmental Mobilization: Trends Among National Environmental Organizations', in R.E. Dunlap and A.G. Mertig (eds), *American Environmentalism: The U.S. Environmental Movement, 1970–1990*, Philadelphia, PA: Taylor & Francis, pp. 11–26.

Mohai, P. and B.W. Twight (1987), 'Age and Environmentalism: An Elaboration of the Buttel Model Using National Survey Evidence', *Social Science Quarterly* **68** (4): 789–815.

Moore, B. (1994), 'Environmental Reinstatement and Compensation: The Insurers' Role, *European Environment* **IV** (1): 24–6.

Morphet, J. (1994), 'Environmental Policy in the EU'. *Local Government Policy Making* **XXI** (1): 26–9.

Mullins, M.L. (1990), 'Statement for the Record', in US House of Representatives, Committee on Energy and Commerce, Subcommittee on Transportation and Hazardous Materials, *Hearings on H.R. 4494: Lender Liability Under Superfund*, 2 August, Washington, DC: USGPO, pp. 176–8.

Mumme, S.P. (1993), 'Environmentalists, NAFTA, and North American Environmental Management', *Journal of Environment and Development* **II** (1): 205–19.

Mundy, D. (1992a), 'Stigma and value', *The Appraisal Journal*, January: 7–13.

Mundy, D. (1992b), 'The Impact of Hazardous Materials on Property Value', *The Appraisal Journal*, April: 155–62.

Nabarro Nathanson, Ltd., Environment Department (1993), 'Environmental Liability — The Europe-Wide Debate, *Environmental Law Matters*, **5** (June): 1–2.

National Research Council (1991), *Environmental Epidemiology*, Washington, DC: National Academy Press.

Needham, B., P. Koenders and B. Kruijt (1993), *Urban Land and Property Markets in The Netherlands*, London: UCL Press.

Needham, B. and B. Kruijt (1992), 'The Netherlands', in B. Wood and R.H. Williams (eds), *Industrial Property Markets in Western Europe*, London: E&FN Spon, Chapter 7.

Needham, D. (1983), *The Economics and Politics of Regulation: A Behavioral Approach*, Boston: Little, Brown & Company.

New York Times (1991), 'U.S. health aide says he erred on Times Beach', 26 May: 20.

Nugent, N. (1991), *The Government and Politics of the European Community*, Second Edition, London: Macmillan.

O'Brien, J.P. (1989), 'EPA's landowner liability guidance', *Toxics Law Reporter* IV (7): 184–91.

Official Journal (1985), 'Directive on the Assessment of the Effects of certain Public and Private Projects on the Environment (Directive EEC/85/337), L 175', 5 July, Luxembourg: Luxembourg: Office of Official Publications of the European Union (OOPEC).

Official Journal (1990), 'Council Regulation on the Establishment of the European Environment Agency (Regulation EEC/90/1210) L 120', 11 May, Luxembourg: OOPEC.

Official Journal (1991a), 'Proposal for a Council Directive on Landfill Waste C 190', 22 July, Luxembourg: OOPEC.

Official Journal (1991b), 'Amended Proposal for a Council Directive on Civil Liability for Damage Caused by Waste C 19', 23 July, Luxembourg: OOPEC.

Organization for Economic Cooperation and Development (OECD) (1989), *Accidents Involving Hazardous Substances*, Environment Monograph No. 24, Paris: OECD.

OECD (1990), *Environmental Policies for Cities in the 1990s*, Paris: OECD.

O'Riordan, T. and J. Cameron (1994), *Interpreting the Precautionary Principle: A Principle for Action in the Face of Uncertainty*, London: Cameron May.

Page, G.W. and H.Z. Rabinowitz (1994), 'Potential for redevelopment of contaminate brownfield sites', *Economic Development Quarterly* VIII (4): 353–63.

Pearson, C.S. (1987), 'Environmental Standards, Industrial Location, and Pollution Havens', in C.S. Pearson (ed.), *Multinational Corporations, Environment, and the Third World*, Durham, NC: Duke University Press, Chapter 5, pp. 113–28.

Peiser, R. and C. Taylor (1994), 'Does the land market break down for contaminated properties?', *Journal of Property Research* 11: 145–58.

Picou, J.S. and D. Rosebrook (1993), 'Technological Accidents, Community Class-Action Litigation, and Scientific Damage Assessment: A Case Study of Court-Ordered Research', *Sociological Spectrum* XIII: 117–38.

Piller, C. (1991), *The Fail-Safe Society: Community Defiance and the End of American Technological Optimism*, Berkeley: University of California Press.

Plant, J. (ed.) (1989), *Healing the Wounds: The Promise of Ecofeminism*, Santa Cruz, CA: New Society Publishers.

Platt, R.H., R.A. Rowntree and P.C. Muick (eds) (1994), *The Ecological City: Preserving and Restoring Urban Biodiversity*, Amherst, MA: University of Massachusetts Press.

Pohl, I. and C. Grüssen (1991), 'Land Reclamation in the Federal Republic of Germany', in *European Environment Yearbook*, London: DocTer International, pp. 279–82.

Pollard, A.M. (1992), 'Commentary on J.M. Campbell's "Lender Liability for Environmental Cleanup: Effect on the Financial Services Industry"', in

References

American Council for Capital Formation, Center for Policy Research (eds), *U.S. Waste Management Policies: Impact on Economic Growth and Investment Strategies*, Washington, DC: American Council for Capital Formation, Center for Policy Research, pp. 67–69.

Portney, P.R. (1992), 'The Economics of Hazardous Waste Regulation', in American Council for Capital Formation, Center for Policy Research (eds), *U.S. Waste Management Policies: Impact on Economic Growth and Investment Strategies*, Washington, DC: American Council for Capital Formation, pp. 1–22.

Probst, K.N. and P.R. Portney (1992), *Assigning Liability for Superfund Cleanups: An Analysis of Policy Options*, Washington, DC: Resources for the Future.

Ramsay, A. (1991), *EUROJARGON: A Dictionary of EU Acronyms, Abbreviations, and Sobriquets*, Stamford, Lincolnshire, UK: Capital Planning Information.

Reilly, W.K. (1993), 'The Greening of NAFTA: Implications for Continental Environmental Cooperation in North America', *Journal of Environment and Development* **II** (1): 181–91.

Renn, O. (1991), 'Strategies of Risk Communication: Observations from Two Participatory Experiments', in R.E. Kasperson and P.J.M. Stallen (eds), *Communicating Risk to the Public*, Boston: Kluwer, pp. 457–81.

Renn, O., W.J. Burns, J.X. Kasperson, R.E. Kasperson and P. Slovic (1992), 'The Social Amplification of Risk: Theoretical Foundations and Empirical Applications', *Journal of Social Issues* **XXXXVIII** (4): 137–60.

RETI, see under Association...

Rich, R.C., W.D. Conn and W.L. Owens (1993), '"Indirect Regulation" of Environmental Hazards through the Provision of Information to the Public: The Case of SARA, Title III', *Policy Studies Journal* **XXI**: 16–34.

Ringquist, E.J. (1993), *Environmental Protection at the State Level*, New York: M. E. Sharpe.

Roper Organization (1990), *Roper Reports* **LXXXX** (2): 25.

Royal Town Planning Institute (UK) (1994), 'Paying for Our Past': Memorandum of Observations, London: Royal Town Planning Institute.

Russell, Milton, E. William Colglazier and Mary R. English (1991), *Hazardous Waste Remediation: The Task Ahead*, Knoxville, TN: Waste Management Research and Education Institute, University of Tennessee.

Sandmann, P.M. (1989), 'Hazard versus Outrage: A Conceptual Frame for Describing Public Perception of Risk' in H. Jungermann, R.E. Kasperson and P.M. Wiedemann (eds), *Risk Communication*, Julich, Germany: Nuclear Research Center, pp. 163–8.

Scherr, S.J. (1987), 'Hazardous Exports: US and International Policy Developments', in C.S. Pearson (ed.), *Multinational Corporations, Environment, and the Third World*, Durham, NC: Duke University Press, Chapter 6, pp. 129–48.

Schnapf, L. (1992), 'The EPA's Lender Liability Rule: Panacea or Pitfall?', *The Real Estate Finance Journal*, Summer: 41–7.

Schweitzer, G.E. (1991), *Borrowed Earth, Borrowed Time: Healing America's Chemical Wounds*, New York: Plenum Press.

Scott, D. and F.K. Willits (1994), 'Environmental Attitudes and Behavior: A Pennsylvania Survey.' *Environment and Behavior* **XXVI** (2): 239–60.

Segerson, K. (1992), 'Lender Liability for Hazardous Waste Cleanup', in T.H. Tietenberg (ed.), *Innovation in Environmental Policy*, Cheltenham, England: Edward Elgar Publishing, Ltd., pp. 195–212.

Seidman, L.W. (1991), 'Prepared Statement.' in United States Congress, Senate Committee on Banking, Housing, and Urban Affairs, *Hearings on Environmental Lender Liability Issues*, 12 June, Washington, DC: USGPO, pp. 53–80.

Setterberg, F. and L. Shavelson, (1993), *Toxic Nation: The Fight to Save Our Communities from Chemical Contamination*, New York: John Wiley & Sons.

Shelbourn, C. (1994), 'Historic Pollution — Does the Polluter Pay?', *Journal of Planning and Environmental Law*, August: 703–9.

Shlay, A. (1993), 'Shaping Place: Institutions and Metropolitan Development Patterns', *Journal of Urban Affairs* **15** (6): 395–412.

Short, J.F. (1984), 'The Social Fabric of Risk: Toward the Social Transformation of Risk Analysis', *American Sociological Review* **IX**: 711–25.

Simmons, Harris H. (1991), 'Statement of the American Bankers Association on Lender Liability Under Superfund', in United States Congress, Senate Committee on Banking, Housing, and Urban Affairs, *Hearings on Environmental Lender Liability Issues*, 12 June, Washington, DC: USGPO, pp. 208–41.

Simons, R.A. and A. Sementelli (1994), 'Management Issues for Leaky Underground Storage Tanks: How Are Property Transactions and Sales Prices Affected by Regulation of Contamination'? Paper presented at the Association of Collegiate Schools of Planning, Phoenix, Arizona, October.

Singh, K. (1994), *Managing Common Pool Resources*, Delhi: Oxford University Press.

Slovic, P. (1979), 'Rating the Risks', *Environment* **XXI**: 14–20, 36–9.

Slovic, P. (1987), 'Perception of Risk', *Science* **CCXXXVI**: 280–85.

Slovic, P. (1992), 'Perception of Risk: Reflections on the Psychometric Paradigm', in S. Krimsky and D. Golding (eds), *Social Theories of Risk*, Westport CT: Praeger, pp. 117–52.

Slovic, P., B. Fischhoff and S. Lichtenstein (1976), 'Cognitive Processes and Societal Risk Taking', in J.S. Carroll and J.W. Payne (eds), *Cognition and Social Behavior*, Potomac, MD: Erlbaum, pp. 165–84.

Smith, O.T. (1992), 'EPA's Regulatory Fix for Lenders', *Real Estate Review*, Spring: 12–14.

Starr, C. (1969), 'Social Benefit versus Technological Risk', *Science* **CLXV**: 1232–8.

Strayer, D. (1992), Presentation to Brownfields Reuse Strategies: A Local Development Symposium, 10 October, Cited in Cuyahoga County (Ohio) Planning Commission. 1992. *Summary of the Proceedings*, Cleveland, OH: Cuyahoga County Planning Commission, p. 29.

Swartz, R.D. (1994), 'Michigan's Approach to Urban Redevelopment Involving Contaminated Properties', *Economic Development Quarterly* **VIII** (4): 329–37.

Tibbetts, J. (1995), 'Everybody's Taking the Fifth', *Planning* **LXI** (1): 4–9.

Tomsho, R. (1991), 'Pollution Ploy: Big Corporations Hit by Superfund Cases Find Way to Share Bill', *Wall Street Journal*, 2 April: 6.

Toulme, N.V. and D.E. Cloud (1991), 'The Fleet Factors Case: A Wrong Turn for Lender Liability under Superfund', *Wake Forest Law Review* **XXV** (1): 17–21.

Toyka, R. (1993), 'Vorwort', in *Bitterfeld Braunkohlew-Brachen Probleme, Chancen, Visionen*, Bauhaus-Dessau, München: Prestel, pp. 17–20.

Tripp, David R. (1991), 'Wichita Strikes Back at the Blob: Municipal Liability under CERCLA and How One City Solved Ground Water Problems and Rejuvenated its Declining Tax Base', *Toxics Law Reporter* **VI** (4): 130–37.

United Kingdom, Department of the Environment (UK DOE) (1993), *Response to the Communication from the Commission of the European Communities (COM (93) 47 final) Green Paper on Remedying Environmental Damage*, Memorandum, 8 October.

UK, DOE (1994), *Framework for Contaminated Land*, London: Her Majesty's Stationery Office (HMSO).

United Kingdom, Department of Environment and Welsh Office (1994), *Paying for Our Past*, Consultation Paper, London: HMSO.

United Kingdom, House of Lords, Select Committee on the European Communities (1993), *Remedying Environmental Damage — 3rd report*, Session 1993–94, HL Paper 10, London: HMSO.

United Kingdom Environmental Law Association (1993), Submissions to the House of Lords, European Committee — Sub-Committee C, 'The European Commission's Green Paper COM(93) 47 final', London: UK Environmental Law Association memorandum.

United Nations Environment Programme, Industry and Environment/Programme Activity Centre (IE/PAC) (1992), *Hazard Identification and Evaluation in a Local Community*, Technical Report No. 12, Paris: UNEP.

United States Congress, Congressional Budget Office (US CBO) (1985), *Hazardous Waste Management: Recent Changes and Policy Alternatives*, Washington DC: US Government Printing Office (USGPO).

US CBO, (1994), *The Total Costs of Cleaning Up Nonfederal Superfund Sites*, Washington DC: USGPO.

United States Congress, General Accounting Office (US GAO) (1992), *Superfund: Problems With the Completeness and Consistency of Site Cleanup Plans*, Washington, DC: USGPO.

US GAO (1993), *Superfund: Cleanups Nearing Completion Indicate Future Challenges*, Washington, DC: USGPO.

US GAO. (1994a), *Federal Facilities: Agencies Slow to Define the Scope and Cost of Hazardous Waste Site Cleanups*, Washington, DC: USGPO.

US GAO (1994b), *Superfund: EPA's Community Relations Efforts Could Be More Effective*, Washington, DC: USGPO.

US GAO (1994c), *Superfund: EPA Has Opportunities to Increase Recoveries of Costs*, Washington, DC: USGPO.

US GAO (1994d), *Superfund: Further EPA Management Action Is Needed to Reduce Legal Expenses*, Washington, DC: USGPO.

US GAO (1994e), *Superfund: Improved Reviews and Guidance Could Reduce Inconsistencies in Risk Assessments*, Washington, DC: USGPO.

US GAO (1994f), *Toxic Substances Control Act: EPA's Limited Progress in Regulating Toxic Chemicals*, Washington, DC: USGPO.

United States Congress, House of Representatives (US House), Committee on Banking, Finance and Urban Affairs, Subcommittee on Policy Research and Insurance (1991), *Hearings on Lender Liability Under Hazardous Waste Laws*, Washington, DC: USGPO.

United States Congress, House of Representatives, Committee on Small Business (1989), *Hearings on Lender Liability Under Superfund*, Washington, DC: USGPO.

United States Congress, Senate (US Senate), Committee on Banking, Housing, and Urban Affairs (1991), *Environmental Lender Liability Issues: Should Government Agencies and Private Lending Institutions be Forced to Pay Enormous Sums to Clean Up Environmental Contamination that they did not Cause? and S. 651 To Improve the Administration of the Federal Deposit Insurance Corporation, and to make Technical Amendments to the Federal Deposit Insurance Act, and the National Bank Act*, Washington, DC: USGPO.

United States Congress, Senate, Committee on Environment and Public Works, Subcommittee on Superfund, Recycling, and Solid Waste Management (1993), *Hearings on Superfund Reauthorization*, Washington, DC: USGPO.

United States Department of Commerce, Small Business Administration (USSBA) (1989), 'List of Cases Involving SBA Ownership in Property with Hazardous Liability or Potential Liability', in US House, Committee on Small Business, *Hearings: Lender Liability under Superfund*, 3 August, Washington, DC: USGPO, pp. 40–68.

United States Environmental Protection Agency (EPA) (1992), National Oil and Hazardous Substances Pollution Contingency Plan: Lender Liability Under CERCLA, *Federal Register* **LVII** (16): 344, 383–5.

van Breugel, L., R.H. Williams and B. Wood (1993), *Multilingual Dictionary of Real Estate*, London: E&FN Spon.

van der Pligt, J. and J. de Boer (1991), 'Contaminated Soil: Public Reactions, Policy Decisions, and Risk Communication', in R.E. Kasperson and P.J.M. Stallen (eds), *Communicating Risks to the Public*, Dordrecht, NL: Kluwer Academic Publishers, pp. 127–44.

Van Liere, K.D. and R.E. Dunlap (1980), 'The social bases of environmental concern: a review of hypotheses, explanations, and empirical evidence', *Public Opinion Quarterly* **22**: 181–97.

Voskuhl, J. (1993), 'Contaminated by Mistrust: Chemical Cleanup Brings Dayhoit Uncertainty, not Relief', *Louisville Courier Journal*, 18 April: A1.

Vyner, H.M. (1988), *Invisible Trauma*, Lexington, MA: Lexington.

Ward, S.V. (1994), *Planning and Urban Change*, London: Paul Chapman Publishing, Ltd.

Warren, M. (1994), *Contaminated Land: Environmental Liabilities*, Cardiff: Eversheds Phillips & Buck.

Weinstein, A.C. (1993), 'Common Law Nuisance, Lucas and Land Use Regulation', Paper presented at the annual meeting of the Association of Collegiate Schools of Planning, Philadelphia, PA, October.

Whelan, R.P. (1978), *Love Canal: Public Health Time Bomb*, report prepared by the Office of Public Health, State of New York, Albany, NY: Office of Public Health.

Williams, R.H. (1983), 'Land Use Planning, Pollution Control and Environmental Assessment in EC Environment Policy', *Planning Outlook* **XXVI** (2): 54–9.
Williams, R.H. (1986), 'EC environment policy, land use planning and pollution control', *Policy and Politics* **XIV** (1): 93–106.
Williams, R.H. (1989), 'European Spatial Planning Strategies and Environment Policy', in G. Ashworth and P.T. Kivell (eds), *Land Water and Sky: European Environmental Planning*, Groningen: Goepers, pp. 9–18.
Williams, R.H. (1990), 'Supranational Environment Policy and Pollution Control', in D. Pinder (ed.), *Western Europe: Challenge and Change*, London: Belhaven, pp. 195–207.
Williams, R.H. (1991), 'Placing Britain in Europe: Four issues for spatial planning', *Town Planning Review*, **LXII** (3): 331-40.
Williams, R.H. (1992), 'European Spatial Planning and the Cityport System', in B.S. Hoyle and D.A Pinder (eds), *European Port Cities in Transition*, London: Belhaven, Chapter 4, pp. 59–79.
Williams, R.H. (1995a), 'Contaminated Land — a Problem for Europe?' *BrachFlächenRecycling* **III** (94): forthcoming.
Williams, R.H. (1995b), 'European Spatial Strategies and Local Development in Central Europe', *European Spatial Research and Policy* **II** (1): forthcoming.
Williams, R.H., P.B. Meyer and K.R. Yount (1994), 'Contaminated Land: can US experience offer a model for EU policy development', Paper presented at the European Environment Conference, Nottingham, September.
Witkin, J. (1992), 'Lender Liability and Economic Redevelopment', Paper presented at Federal Reserve Bank of Cleveland Conference on 'The Environment and Economic Development in the Great Lakes Region', 24 September.
Wood, B. and R.H. Williams (eds) (1992), *Industrial Property Markets in Western Europe*, London: E&FN Spon.
Wurzel, R. (1993), 'Environmental Policy', in J. Lodge (ed.), *The European Community and the Challenge of the Future*, Second Edition, London: Pinter, pp. 178–99.
Yount, K.R. and P.B. Meyer (1993), 'Fear and Loathing in Urban America: Environmental Concerns and the Prospects for Renewal of Brownfield Sites', *Working Paper 93–3*, Louisville, KY: Center for Environmental Management, University of Louisville.
Yount, K.R. and P.B. Meyer (1994a), 'Who Will Pay for Reclamation of Urban Environmental Blight? Policy Potential in Light of Developer and Lender Risk Perceptions', *Working Paper 94–2*, Louisville, KY: Center for Environmental Management, University of Louisville.
Yount, K.R. and P.B. Meyer (1994b), 'Bankers, Developers, and New Investment in Brownfield Sites: Environmental Concerns and the Social Psychology of Risk', *Economic Development Quarterly* **8** (4): 338–45.
Zeiss, C. and J. Atwater (1989), 'Waste Facility Impacts on Residential Property Values', *Journal of Urban Planning and Development*, **115** (2): 64–80.
Zimmerman, M.E., J.B. Callicott, G. Sessions, K.J. Warren and J. Clark (eds) (1993), *Environmental Philosophy: From Animal Rights to Radical Ecology*, Englewood Cliffs, NJ: Prentice Hall.

Index

abandoned land 3, 4, 18, 27, 48, 85, 124, 132, 149, 160–164, 172
Action Programmes on the Environment, EU 113, 116
anti-regulationist organizations 69, 167
Austria 48, 105, 107, 111, 123, 127
Belgium 28, 29, 48, 105, 111, 123, 126, 129, 137, 138, 164, 167
Bhopal, India 69, 148
Bitterfeld, Germany 29, 48, 109, 141, 179
Britain ix, 3, 4, 9–10, 15, 20, 34, 42, 45, 48, 114–117, 126–127, 130–131, 134
brownfields 4, 18–19, 41, 85–87, 96, 159–164, 174, 191
brownlining 85, 160, 172
Brundtland Commission 112
Bulgaria 108, 123
capital flight 3, 27, 47, 160, 162–164, 182, 198
Carter, President, US 57, 63, 66, 68
CE *see* Council of Europe
CEE *see* Central and Eastern Europe
Central and Eastern Europe 8, 105, 108–109, 201
central standards *see* mandates
centralization of lending 27, 45, 160, 161, 163, 170
CERCLA *see* Comprehensive Environmental Reclamation, Compensation, and Liability Act
Clean Air Act, US 56, 57
cleanup liability 26, 37, 154
cleanup requirements *see* cleanup standards
cleanup standards 22, 27, 31–33, 48, 71, 74, 75, 94, 130, 132, 143, 144
Clinton, President, US 12, 72, 96, 182
collective management 178, 180–181, 200
collective responsibility (for clean-up) 100, 159, 186, 199
common pool resource 16, 17, 175–182, 192, 195, , 197, 199–200
    principles for managing 175–178
community control *see* collective management
Comprehensive Environmental Reclamation, Compensation, and Liability Act, US 9–10, 12, 14, 17, 19, 47, 49–50, 55, 57, 60–61, 63–67, 71–92, 95–100, 112, 128, 132, 134, 136, 143, 144, 147–148, 152, 154, 156–158, 161–162, 165, 172, 173, 176, 177, 179, 182–188, 193–198, 201
    early history 64–67
    financing the Fund 76–77
    hazards ranking system (HRS) 73–74
    impact on redevelopment 84–91
    liability provisions under 77–79, 158, 160, 172

Index

National Priority List (NPL) 65, 71–75, 72, 77, 79, 87–89, 91, 92, 94, 96–99, 171, 179–180, 182, 193
  cleanup problems 71–75
  successes 10, 12, 51, 71, 98, 158, 176–178, 197
  Superfund 46, 71, 76, 77, 97, 98, 139, 186
contaminated land policy 6, 7, 11, 12, 24, 26, 43, 48, 96, 119–121, 127–129, 137, 147, 166–170, 172, 175–178, 181–183, 186, 192, 197, 199
  and nationalization 178–180, 199
  and privatization 173, 178–180, 186, 199
  corporatism 150, 173, 198, 199
  decentralized 178, 180, 181, 200
  in EU 121, 132, 152
  in EU member-states 137–144
  in US see Comprehensive Environmental Reclamation, Compensation, and Liability Act
contaminated land problem 3, 5–6, 11, 13–16, 51, 67, 103, 121, 122, 132
  competitive position of contaminated lands 169–175
  extent of 11, 13–15, 104, 122–125
  parties affected by 11–15, 189–195
  public concern and 61–64, 124–126, 165–168, 189–193
    in the EU 124–126
    in the US 61–64
  risk perception and 175, 191–193
corporatist values and policies 29, 47, 135, 149, 151, 155, 159, 179, 202
  elements of effective policy 181–187
cost-sharing (for clean-up) 22, 23, 26, 29, 40, 44, 47, 165, 197
Council of Europe 19, 104, 129
Council of Ministers, EU 110, 115, 127, 128
Covenants not to sue 91, 92, 97
CPR see common pool resource
current landowners 8, 84–86, 133–135, 186
Czech Republic 108, 123
'deep pockets' 38, 144
Defenders of Wildlife, US 58
Denmark 15, 105, 111, 115, 123, 124, 126, 133–135
Department of Environment, UK 10, 20, 112, 119, 123, 130, 131, 133, 134, 143, 144
Derelict Land Grant Programme 4
dereliction see abandoned land
development limits 26, 35, 151, 153
Directorate-General for Environment, EU 115
disclosure 26, 38, 39, 92, 154, 156
DOE see Department of Environment
EEA see European Environment Agency
EEC see European Economic Community
EFTA see European Free Trade Area
Endangered Species Act, US 56, 57
Environmental Action, US 58, 113, 126, 166

Environmental Action Programme, UK 126, 166
Environmental Assessment Directive, EU 105, 112
Environmental Defense Fund, US 58, 70
environmental justice 31, 96, 166, 197
environmental lobbies 167
environmental movements 25, 27, 50, 124, 126, 167
Environmental Pesticide Control Act, US 57
Environmental Policy Institute, US 58
Environmental Protection Act, UK 14, 142, 156
Environmental Protection Agency, US 5, 10, 55, 57–70, 73–84, 87, 91, 92, 95–100, 147, 157–158, 168, 176
   description of 58–60
environmental racism *see* environmental justice
environmentalism 50, 55, 57, 67, 68, 168
EPA *see* Environmental Protection Agency
EU *see* European Union
European Commission 20, 105–108, 110, 112, 114–117, 121, 125, 128, 129, 131, 132, 134, 144
European Community 110, 115, 126
European Economic Area 8, 48, 108
European Economic Community 104–107, 113, 115, 120
European Environment Agency 105, 107, 108, 110, 127, 168, 172, 195
European Free Trade Area 105, 108, 109
European Parliament 106, 110, 114, 115, 127, 131, 167
European Union 3, 8, 10–14, 19, 20, 22, 23, 27–29, 48, 49, 51, 103–117, 119–138, 141, 142, 144, 147–155, 157, 159, 160, 162, 164, 166–169, 170, 172–176, 178, 179, 181, 182, 184, 186, 187, 193–196, 198–201
Federal Insecticide, Fungicide, and Rodenticide Act, US 57, 60
Federal Land Policy and Management Act, US 57, 83
federalism 10, 11, 29, 137
financial responsibility (for contamination or cleanup) 7, 8, 37, 43, 44, 191, 196
Finland 48, 105, 107, 109, 111, 123, 127
four freedoms, EU 108, 113, 114, 197
France 4, 12, 13, 15, 29, 48, 105, 111, 123, 125, 126, 133, 134, 138, 167
Friends of the Earth 58, 125
future liability 26, 39, 79, 80, 97, 98–152, 154, 158–159, 171, 186, 201
GAO *see* Government Accounting Office
garden festivals 179
German Democratic Republic (GDR) 29, 48, 108, 112, 122, 136, 141, 164
Germany 9, 10, 12, 15, 20, 29, 48, 105, 108, 109, 111, 112, 119, 122–128, 130, 132–134, 136–140, 144, 149, 164, 167, 179, 187
Gorsuch-Buford, Anne 66, 67
Government Accounting Office, US 64, 72–75, 77, 79, 95, 179
Greece 48, 105, 109, 111, 123, 126, 128
Green Paper on Remedying Environmental Damage, CEC 20, 105, 129, 131, 132, 134
Green Paper on the Urban Environment, CEC 116, 119, 131

Index   221

Green parties 27, 50, 113, 125–127, 167
Greenpeace 125
Greens *see* Green parties
Hazardous substance regulation 60–61, 105, 148
Hungary 108, 123
Iceland 108
individualist values and policies 47, 100, 149–151, 156, 173, 199, 201
  pressures for change 194–199
insurance 23, 27, 44–47, 60, 84, 90, 98, 119, 124, 159, 162, 163, 194
Ireland 33, 105, 111, 118, 123, 126, 128, 152
Italy 25, 48, 50, 105, 106, 111, 123, 124, 126, 134, 149, 179
Izaak Walton League, US 58
land markets 118, 119, 125, 149, 200
land reclamation 3, 48, 92, 94, 130, 132, 135, 138, 141, 142, 144, 198
land use controls 33–37, 151–152
Lavelle, Rita 66, 67, 69
lending practices 27, 44, 160, 162, 163
liability 7, 9, 10, 17, 19–23, 25–27, 29, 36–39, 43, 44, 47, 50, 55, 66, 70, 71,
    77–88, 90–93, 96–100, 105, 119, 124, 125, 128–131, 133–136, 139, 140,
    142–144, 152, 154–164, 169–172, 180, 182, 184, 186, 187, 189, 190,
    193, 201
  allocation of 26, 36, 37, 50, 154–156
  joint and several 37, 38, 66, 77, 128, 134, 154, 157, 158, 172, 180
  of lenders 79, 80, 82, 96, 170
  private 77–82, 154, 155, 184
  strict 29, 31, 37, 38, 66, 77, 128, 133–136, 143, 144, 149, 154, 156–160,
    166, 169, 172, 182, 201
  under CERCLA 50, 55, 71, 74, 77, 79, 80, 83–85, 88, 132, 143, 152, 158,
    161, 172, 173, 184, 185, 188, 193
Lichtenstein 192, 193
local state 27, 30, 34, 155, 165, 180, 182
Love Canal, US 25, 62, 63, 65, 139, 147
Lugano Convention, CE 19, 20, 172
Luxembourg 105, 111, 123, 126
Maastricht Treaty *see* Treaty of European Union
mandates 26–30, 75, 147, 150, 153
NAFTA *see* North American Free Trade Agreement
National Audubon Society, US 58
National Energy Act 57
National Parks and Conservation Association, US 58
National Priority List (NPL) *see* Comprehensive Environmental Reclamation,
    Compensation, and Liability Act
National Wildlife Federation, US 58
Natural Resources Defense Council, US 58
negligence 77, 134
Netherlands 4, 15, 20, 29, 105, 111, 118, 123–126, 130, 133–135, 144, 149,
    150, 172, 179
new landowners 26, 38, 86, 154–158

new owner 86, 157
Nixon, President, US 57, 67
North American Free Trade Agreement 11, 28, 48, 103, 104, 147, 174, 196, 199
North American Free Trade Area 8
Norway 108
Ocean Dumping Act, US 57
'open-cast' mines 122
open pit mines 123
PHARE 108, 109
Poland 108, 123
polluter pays principle 17, 20, 65, 76, 77, 106, 112, 131, 143, 159, 183–185, 199, 201
Portugal 105, 111, 123, 126, 128, 134
Potentially responsible parties (PRP's, under CERCLA) 36, 76, 82, 152, 156, 157, 165, 173, 179, 180, 201
Precautionary principle, EU 106, 112, 144
pressures for policy 176, 192, 194, 195
private redevelopment 131, 152, 162
property rights 16, 24, 26, 32, 34, 91, 152, 153
prospective liability *see* future liability
Reagan, President, US 66–69, 72, 176
Resource Conservation and Recovery Act, US 57, 60
retrospective liability 143, 184
return on investment 21, 157
risk aversion 27, 45, 46, 163, 175, 189, 193
risk exposure 46, 157, 161, 187, 190
Romania 108, 123
Ruhrgebiet, Germany 48, 50, 129, 138–140, 179
Safe Drinking Water Act, US 57, 60
SARA *see* Superfund Amendment and Reauthorization Act
SEM *see* Single European Market
Seveso 25, 124
Seveso Directive 105, 148, 195
Sierra Club, US 58, 68
Single European Act 105–106
Single European Market 28, 105–106, 108, 114, 115, 117, 126–127, 130, 135–136, 147, 168, 173, 196–197
site mitigation *see* land reclamation
site-specific approach to cleanups 149, 181
Slovakia 108, 123
Soviet Union 108, 109
Spain 29, 48, 105, 111, 123, 125, 126
spatial policy 26, 30, 150
state responsibility 29, 31, 39, 139, 154, 155, 171, 173, 179, 186, 187
stigmatization (of property) 15, 21, 22, 89, 90, 143, 170, 172
'suitable-for-use' principle *see* cleanup standards

subsidiarity (principle in EU) 106, 107, 112, 113, 116, 119, 130–133, 142, 149, 150
Superfund *see* Comprehensive Environmental Reclamation, Compensation, and Liability Act
Superfund Amendment and Reauthorization Act 9, 55, 67, 69, 70, 74–77, 79, 85, 97, 148, 195
Superfund Reauthorization Act 72, 75
  provisions of 95–98
Surface Mining Control and Reclamation Act, US 57
sustainability 105, 112, 116, 167, 175, 178
Sweden 48, 105, 107, 111, 123, 124, 127, 136, 138
Switzerland 29, 108
TACIS Programme 109
taxation 12, 29, 48, 136, 165, 184, 185, 187
Times Beach, US 25, 68, 70
Toxic Substances Control Act, US 57, 60, 158, 177
Transaction costs 19–21, 39, 40, 79, 85–87, 93, 97, 147, 169, 170, 185, 186, 198
Treaty of European Union 105, 106, 110–112
  Environment Title 106, 111, 144
Treaty of Rome 104, 105
TSCA *see* Toxic Substances Control Act
United Kingdom (UK) 3, 10, 14, 20, 21, 33, 105, 112–114, 118, 119, 122–126, 128, 130–135, 137, 138, 141–144, 149, 150, 152, 154–156, 161, 164, 167, 170, 172, 173, 176, 179, 184, 186
Urban Development Corporations (UDCs), UK 42, 149
voluntary risk takers 187, 189
Water Pollution Control Act, US 57
White Paper on Growth, Competition and Employment, CEC 132
Wilderness Act, US 56
Wilderness Society, US 58, 68
zoning 22, 26, 32–34, 94, 95, 152, 153, 165, 169